**Cambridge
International AS & A Level**

Further
Mathematics
Further Mechanics

Jean-Paul Muscat and
Sophie Goldie
Series editor: Roger Porkess

Questions from the Cambridge International AS & A Level Further Mathematics papers are reproduced by permission of Cambridge Assessment International Education. Unless otherwise acknowledged, the questions, example answers, and comments that appear in this book were written by the authors. Cambridge Assessment International Education bears no responsibility for the example answers to questions taken from its past question papers which are contained in this publication.

The publishers would like to thank the following who have given permission to reproduce photographs in this book:

p.1 © photo7/123RF; **p.28** © Anton Havelaar/123RF; **p.56** © alphaspirit/123RF; **p.75** Photo courtesy of Millbrook Proving Ground Ltd; **p.86** *l* © M.Rosenwirth - Fotolia, *c* © Michael Steele/Getty Images, *r* © NickR - Fotolia; **p.99** © Steeve ROCHE - Fotolia; **p.116** © Colin Anderson - Getty Images; **p.127** *l* © elina - Shutterstock, *c* © ChameleonsEye - Shutterstock, *r* © ArtemZ - Shutterstock; **p.141** *t* © Aggie 11 - Shutterstock, *b* © Matthew Ashmore/Alamy Stock Photo; **p.155** © mulderphoto/adobe.stock.com; **p.167** © Blend Images/Alamy Stock PhotoJGI/Tom Grill. *l* = left, *c* = centre, *r* = right, *t* = top, *b* = bottom

Every effort has been made to trace and acknowledge ownership of copyright. The publishers will be glad to make suitable arrangements with any copyright holders whom it has not been possible to contact.

Hachette UK's policy is to use papers that are natural, renewable and recyclable products and made from wood grown in well-managed forests and other controlled sources. The logging and manufacturing processes are expected to conform to the environmental regulations of the country of origin.

Orders: please contact Hachette UK Distribution, Hely Hutchinson Centre, Milton Road, Didcot, Oxfordshire, OX11 7HH. Telephone: +44 (0)1235 827827. Email education@hachette.co.uk Lines are open from 9 a.m. to 5 p.m., Monday to Friday. You can also order through our website: www.hoddereducation.com.

Much of the material in this book was published originally as part of the MEI Structured Mathematics series. It has been carefully adapted to support the Cambridge International AS & A Level Further Mathematics syllabus. The original MEI author team for Mechanics comprised John Berry, Pat Bryden, Ted Graham, Cliff Pavelin and Roger Porkess.

First published in 2018 by

Hodder Education, an Hachette UK company,

Carmelite House, 50 Victoria Embankment,

London EC4Y 0DZ

Impression number 5 4

Year   2022

Cover photo by Shutterstock/Arm Emer

Illustrations by Pantek Media and Integra Software Services

Typeset in Bembo Std 11/13 Integra Software Services Pvt Ltd, Pondicherry, India

Printed and bound by CPI Group (UK) Ltd, Croydon, CR0 4YY

A catalogue record for this title is available from the British Library

ISBN   9781510421806

Contents

Introduction

This is one of a series of four books supporting the Cambridge International AS & A Level Further Mathematics 9231 syllabus for examination from 2020. It is preceded by five books supporting Cambridge International AS & A Level Mathematics 9709. The seven chapters in this book cover the further mechanics required for the Paper 3 examination. This part of the series also contains two books for further pure mathematics and one book for further probability and statistics.

These books are based on the highly successful series for the Mathematics in Education and Industry (MEI) syllabus in the UK but they have been redesigned and revised for Cambridge International students; where appropriate, new material has been written and the exercises contain many past Cambridge International examination questions. An overview of the units making up the Cambridge International syllabus is given in the following pages.

Throughout the series, the emphasis is on understanding the mathematics as well as routine calculations. The various exercises provide plenty of scope for practising basic techniques; they also contain many typical examination-style questions.

The original MEI author team would like to thank Jean-Paul Muscat and Sophie Goldie who have carried out the extensive task of presenting their work in a suitable form for Cambridge International students and for their many original contributions. They would also like to thank Cambridge Assessment International Education for its detailed advice in preparing the books and for permission to use many past examination questions.

Roger Porkess

Series editor

How to use this book

The structure of the book

This book has been endorsed by Cambridge Assessment International Education. It is listed as an endorsed textbook for students taking the Cambridge International AS & A Level Further Mathematics 9231 syllabus. The Further Mechanics syllabus content is covered comprehensively and is presented across seven chapters, offering a structured route through the course.

The book is written on the assumption that you have covered and understood the work in the Cambridge International AS & A Level Mathematics 9709 syllabus, including the mechanics content. The following icon is used to indicate material that is not directly on the syllabus:

> **e** There are places where the book goes beyond the requirements of the syllabus to show how the ideas can be taken further or where fundamental underpinning work is explored. Such work is marked as **extension**.

Each chapter is broken down into several sections, with each section covering a single topic. Topics are introduced through **explanations**, with **key terms** picked out in red. These are reinforced with plentiful **worked examples**, punctuated with commentary, to demonstrate methods and illustrate application of the mathematics under discussion.

Regular **exercises** allow you to apply what you have learned. They offer a large variety of practice and higher-order question types that map to the key concepts of the Cambridge International syllabus. Look out for the following icons.

PS **Problem-solving questions** will help you to develop the ability to analyse problems, recognise how to represent different situations mathematically, identify and interpret relevant information, and select appropriate methods.

M **Modelling questions** provide you with an introduction to the important skill of mathematical modelling. In this, you take an everyday or workplace situation, or one that arises in your other subjects, and present it in a form that allows you to apply mathematics to it.

CP **Communication and proof questions** encourage you to become a more fluent mathematician, giving you scope to communicate your work with clear, logical arguments and to justify your results.

Exercises also include questions from real Cambridge Assessment International Education past papers, so that you can become familiar with the types of questions you are likely to meet in formal assessments.

Answers to exercise questions, excluding long explanations and proofs, are available online at www.hoddereducation.com/cambridgeextras, so you can check your work. It is important, however, that you have a go at answering the questions before looking up the answers if you are to understand the mathematics fully.

In addition to the exercises, a range of additional features are included to enhance your learning.

> ▶ **ACTIVITY**
>
> **Activities** invite you to do some work for yourself, typically to introduce you to ideas that are then going to be taken further. In some places, activities are also used to follow up work that has just been covered.

EXPERIMENT

In applied mathematics (mechanics and statistics), it is often helpful to carry out **experiments** so that you can see for yourself what is going on. The same is sometimes true for pure mathematics, where a spreadsheet can be a particularly powerful tool.

Other helpful features include the following.

 This symbol highlights points it will benefit you to **discuss** with your teacher or fellow students, to encourage deeper exploration and mathematical communication. If you are working on your own, there are answers available online at www.hoddereducation.com/cambridgeextras.

 This is a **warning** sign. It is used where a common mistake, misunderstanding or tricky point is being described to prevent you from making the same error.

A variety of notes are included to offer advice or spark your interest:

> **Note**
>
> **Notes** expand on the topic under consideration and explore the deeper lessons that emerge from what has just been done.

> **Historical note**
>
> **Historical notes** offer interesting background information about famous mathematicians or results to engage you in this fascinating field.

Finally, each chapter ends with the **key points** covered, plus a list of the **learning outcomes** that summarise what you have learned in a form that is closely related to the syllabus.

Digital support

Comprehensive online support for this book, including further questions, is available by subscription to MEI's Integral® online teaching and learning platform for AS & A Level Mathematics and Further Mathematics, integralmaths.org. This online platform provides extensive, high-quality resources, including printable materials, innovative interactive activities, and formative and summative assessments. Our eTextbooks link seamlessly with Integral, allowing you to move with ease between corresponding topics in the eTextbooks and Integral.

MEI's Integral® material has not been through the Cambridge International endorsement process.

The Cambridge International AS & A Level Further Mathematics 9231 syllabus

The syllabus content is assessed over four examination papers.

Paper 1: Further Pure Mathematics 1	Paper 3: Further Mechanics
• 2 hours	• 1 hour 30 minutes
• 60% of the AS Level; 30% of the A Level	• 40% of the AS Level; 20% of the A Level
• Compulsory for AS and A Level	• Offered as part of AS; compulsory for A Level
Paper 2: Further Pure Mathematics 2	**Paper 4: Further Probability & Statistics**
• 2 hours	• 1 hour 30 minutes
• 30% of the A Level	• 40% of the AS Level; 20% of the A Level
• Compulsory for A Level; not a route to AS Level	• Offered as part of AS; compulsory for A Level

The following diagram illustrates the permitted combinations for AS Level and A Level.

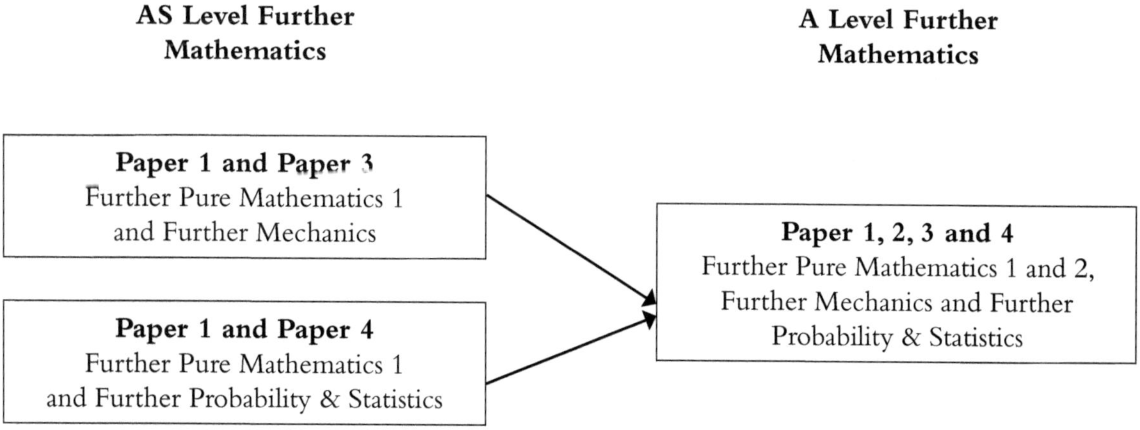

AS Level Further Mathematics

A Level Further Mathematics

Paper 1 and Paper 3
Further Pure Mathematics 1 and Further Mechanics

Paper 1 and Paper 4
Further Pure Mathematics 1 and Further Probability & Statistics

Paper 1, 2, 3 and 4
Further Pure Mathematics 1 and 2, Further Mechanics and Further Probability & Statistics

Prior knowledge

It is expected that learners will have studied the majority of the Cambridge International AS & A Level Mathematics 9709 syllabus content before studying Cambridge International AS & A Level Further Mathematics 9231.

The prior knowledge required for each Further Mathematics component is shown in the following table.

Component in AS & A Level Further Mathematics 9231	Prior knowledge required from AS & A Level Mathematics 9709
9231 Paper 1: Further Pure Mathematics 1	9709 Papers 1 and 3
9231 Paper 2: Further Pure Mathematics 2	9709 Papers 1 and 3
9231 Paper 3: Further Mechanics	9709 Papers 1, 3 and 4
9231 Paper 4: Further Probability & Statistics	9709 Papers 1, 3, 5 and 6

For Paper 3: Further Mechanics, knowledge of Cambridge International AS & A Level Mathematics 9709 Paper 4: Mechanics syllabus content is assumed.

Command words

The table below includes command words used in the assessment for this syllabus. The use of the command word will relate to the subject context.

Command word	What it means
Calculate	work out from given facts, figures or information
Deduce	conclude from available information
Derive	obtain something (expression/equation/value) from another by a sequence of logical steps
Describe	state the points of a topic / give characteristics and main features
Determine	establish with certainty
Evaluate	judge or calculate the quality, importance, amount or value of something
Explain	set out purposes or reasons / make the relationships between things evident / provide why and/or how and support with relevant evidence
Identify	name/select/recognise
Interpret	identify meaning or significance in relation to the context
Justify	support a case with evidence/argument
Prove	confirm the truth of the given statement using a chain of logical mathematical reasoning
Show (that)	provide structured evidence that leads to a given result
Sketch	make a simple freehand drawing showing the key features
State	express in clear terms
Verify	confirm a given statement/result is true

Key concepts

Key concepts are essential ideas that help students develop a deep understanding of mathematics.

The key concepts are:

Problem solving

Mathematics is fundamentally problem solving and representing systems and models in different ways. These include:

» Algebra: this is an essential tool which supports and expresses mathematical reasoning and provides a means to generalise across a number of contexts.

» Geometrical techniques: algebraic representations also describe a spatial relationship, which gives us a new way to understand a situation.

» Calculus: this is a fundamental element which describes change in dynamic situations and underlines the links between functions and graphs.

» Mechanical models: these explain and predict how particles and objects move or remain stable under the influence of forces.

» Statistical methods: these are used to quantify and model aspects of the world around us. Probability theory predicts how chance events might proceed, and whether assumptions about chance are justified by evidence.

Communication

Mathematical proof and reasoning is expressed using algebra and notation so that others can follow each line of reasoning and confirm its completeness and accuracy. Mathematical notation is universal. Each solution is structured, but proof and problem solving also invite creative and original thinking.

Mathematical modelling

Mathematical modelling can be applied to many different situations and problems, leading to predictions and solutions. A variety of mathematical content areas and techniques may be required to create the model. Once the model has been created and applied, the results can be interpreted to give predictions and information about the real world.

These key concepts are reinforced in the different question types included in this book: **Problem-solving**, **Communication and proof**, and **Modelling**.

1

Motion of a projectile

Swift of foot
was Hiawatha;
He could shoot
an arrow from
him, And run
forward with
such fleetness,
That the arrow
fell behind
him! Strong
of arm was
Hiawatha;
He could shoot
ten arrows
upwards,
Shoot them
with such
strength and
swiftness,
That the last
had left the
bowstring,
Ere the first
to earth had
fallen!
*The Song of
Hiawatha,
Longfellow
(1807–1882)*

Look at the water jets in the photograph. Every drop of water in a water jet follows its own path, which is called its **trajectory**. You can see the same sort of trajectory if you throw a small object across a room. Its path is a parabola. Objects moving through the air like this are called **projectiles**.

1.1 Modelling assumptions for projectile motion

The path of a cricket ball looks parabolic, but what about a boomerang? There are modelling assumptions that must be satisfied for the motion to be parabolic. These are:

- » a projectile is a particle
- » it is not powered
- » the air has no effect on its motion.

Equations for projectile motion

A projectile moves in two dimensions under the action of only one force, the force of gravity, which is constant and acts vertically downwards. This means that the acceleration of the projectile is $g\,\mathrm{m\,s^{-2}}$ vertically downwards and there is no horizontal acceleration. You can treat the horizontal and vertical motions separately, using the equations for constant acceleration.

Horizontal distance travelled is small enough to assume that gravity is always in the same direction.	Vertical distance travelled is small enough to assume that gravity is constant.

> **Note**
>
> The value of g varies around the world, from 9.766 in Kuala Lumpur to 9.825 in Oslo. A value of 10 is used in this book.

> **Note**
>
> You have already worked with vectors in *Pure Mathematics 3*. In this chapter they are used to make it easier to distinguish between motion in the horizontal and vertical directions. Although the Cambridge International syllabus does not require students to use vector methods, vectors can provide a useful way to simplify and solve mechanics problems.

To illustrate the ideas involved, think of a ball being projected with a speed of $20\,\mathrm{m\,s^{-1}}$ at 60° to the ground, as illustrated in Figure 1.1. This could be a first model for a football, a chip shot from the rough at golf or a lofted shot at cricket.

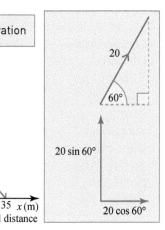

▲ **Figure 1.1**

Using axes as shown, the components are:

	Horizontal	Vertical
Initial position	0	0
Acceleration	$a_x = 0$	$a_y = -10$

This is negative because the positive y-axis is upwards. As a vector
$$\mathbf{a} = \begin{pmatrix} 0 \\ -10 \end{pmatrix}$$

Initial velocity	$u_x = 20\cos 60°$	$u_y = 20\sin 60°$
	$= 10$	$= 17.3\ldots$

As a vector
$$\mathbf{u} = \begin{pmatrix} 20\cos 60° \\ 20\sin 60° \end{pmatrix}$$

'17.3...' means 17.3 and subsequent figures. Keep this number in your calculator for use in future working. You should not round values that will be used in later calculations. Convention for this course is to give final answers to three significant figures (or one decimal place for angles in degrees), unless a question asks for something different.

Using $v = u + at$ in the two directions gives the components of the velocity.

Instead of writing
17.3... every
time, here the
number is written
rounded to four
significant figures
for neatness.
This is one more
significant figure
than required
for the final
answer. However,
remember to use
the value on your
calculator 17.3...
in calculations.

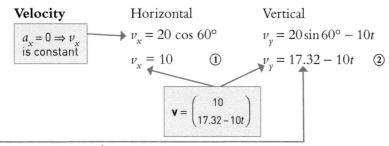

Velocity　　　　　Horizontal　　　　　Vertical

$a_x = 0 \Rightarrow v_x$ is constant

$v_x = 20 \cos 60°$　　$v_y = 20 \sin 60° - 10t$

$v_x = 10$ ①　　$v_y = 17.32 - 10t$ ②

$$\mathbf{v} = \begin{pmatrix} 10 \\ 17.32 - 10t \end{pmatrix}$$

Using $s = ut + \frac{1}{2}at^2$ in the two directions gives the components of position.

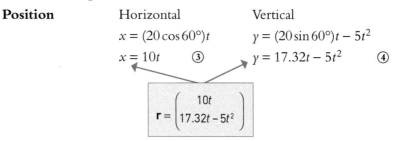

Position　　　　　Horizontal　　　　　Vertical

$x = (20 \cos 60°)t$　　$y = (20 \sin 60°)t - 5t^2$

$x = 10t$ ③　　$y = 17.32t - 5t^2$ ④

$$\mathbf{r} = \begin{pmatrix} 10t \\ 17.32t - 5t^2 \end{pmatrix}$$

You can summarise these results in a table.

	Horizontal motion	**Vertical motion**
initial position	$x_0 = 0$	$y_0 = 0$
a	$a_x = 0$	$a_y = -10$
u	$u_x = 20 \cos 60° = 10$	$u_y = 20 \sin 60° = 17.32$
v	$v_x = 10$ ①	$v_y = 17.32 - 10t$ ②
r	$x = 10t$ ③	$y = 17.32t - 5t^2$ ④

The four equations ①, ②, ③ and ④ for velocity and position can be used to find several things about the motion of the ball.

> What can you say about v_y when the projectile is at the topmost point of its path?
> What can you say about y when the projectile is just about to hit the ground?

When you have decided the answer to these questions you have sufficient information to find the greatest height reached by the ball, the time of flight and the **range** (the total distance travelled horizontally before it hits the ground).

Answers to exercises are available at www.hoddereducation.com/cambridgeextras

The maximum height

When the ball is at its maximum height, H m, the vertical component of its velocity is zero. It still has a horizontal component of $10\,\text{m}\,\text{s}^{-1}$, which is constant.

Equation ② gives the vertical component as

$$v_y = 17.32 - 10t$$

At the top: $0 = 17.3... - 10t$

$$t = \frac{17.3...}{10}$$

$$= 1.73...$$

Remember to use the value on your calculator 17.3... in calculations, not the rounded value 17.32.

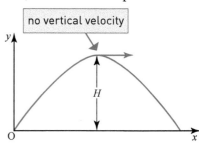

no vertical velocity

▲ Figure 1.2

To find the maximum height, you now need to find y at this time. Substituting for t in equation ④,

$$y = 17.3... \times 1.73... - 5 \times 1.73...^2$$

$$= 15$$

The maximum height is 15 m.

The time of flight

The flight ends when the ball returns to the ground, that is, when $y = 0$. Substituting $y = 0$ in equation ④,

$$y = 17.32t - 5t^2$$

$$0 = 17.32t - 5t^2$$

$$0 = t(17.32 - 5t)$$

$$t = 0 \text{ or } t = 3.46...$$

Clearly $t = 0$ is the time when the ball is projected, so $t = 3.46...$ is the time when it lands and the flight time is 3.46 s.

The range

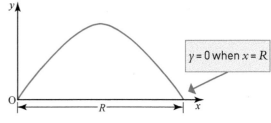

$y = 0$ when $x = R$

▲ Figure 1.3

The range, R m, of the ball is the horizontal distance it travels before landing.

R is the value of x when $y = 0$.
R can be found by substituting $t = 3.46...$ in equation ③ : $x = 10t$.
$R = 10 \times 3.46... = 34.6$ m.

> › Notice in the example of the ball shown in Figure 1.1 that the time to maximum height is half the flight time. Is this always the case?
>
> › Decide which of the following could be modelled as projectiles.
>
> | | | | |
> |---|---|---|---|
> | a balloon | a bird | a glider | a cannonball |
> | a golf ball | a parachutist | a rocket | a tennis ball |
>
> › What special conditions would have to apply in particular cases?

Exercise 1A

In this exercise take upwards as positive. All the projectiles start at the origin.

1 In each case you are given the initial velocity of a projectile.

   (a) Draw a diagram showing the initial velocity and path.

   (b) Write down the horizontal and vertical components of the initial velocity.

   (c) Write down equations for the velocity after time t seconds.

   (d) Write down equations for the position after time t seconds.

   (i) $10\,\mathrm{m\,s^{-1}}$ at 35° above the horizontal.

   (ii) $2\,\mathrm{m\,s^{-1}}$ horizontally, $5\,\mathrm{m\,s^{-1}}$ vertically.

   (iii) $4\,\mathrm{m\,s^{-1}}$ horizontally.

   (iv) $10\,\mathrm{m\,s^{-1}}$ at 13° below the horizontal.

   (v) $U\,\mathrm{m\,s^{-1}}$ at angle α above the horizontal.

   (vi) $u_0\,\mathrm{m\,s^{-1}}$ horizontally, $v_0\,\mathrm{m\,s^{-1}}$ vertically.

2 In each case find

   (a) the time taken for the projectile to reach its highest point

   (b) the maximum height.

   (i) Initial velocity $5\,\mathrm{m\,s^{-1}}$ horizontally and $15\,\mathrm{m\,s^{-1}}$ vertically.

   (ii) Initial velocity $10\,\mathrm{m\,s^{-1}}$ at 30° above the horizontal.

3 In each case find

   (a) the time of flight of the projectile

   (b) the horizontal range.

   (i) Initial velocity $20\,\mathrm{m\,s^{-1}}$ horizontally and $20\,\mathrm{m\,s^{-1}}$ vertically.

   (ii) Initial velocity $5\,\mathrm{m\,s^{-1}}$ at 60° above the horizontal.

CP 4 A ball is projected from ground level with initial velocity $\mathbf{u} = \begin{pmatrix} u_x \\ u_y \end{pmatrix}$. In terms of u_x, u_y and g, find

   (i) the maximum height

   (ii) the time of flight

   (iii) the range.

Answers to exercises are available at www.hoddereducation.com/cambridgeextras

1.2 Projectile problems

Representing projectile motion by vectors

Figure 1.4 shows a possible path for a marble that is thrown across a room, from the moment it leaves the hand until the instant at which it hits the floor.

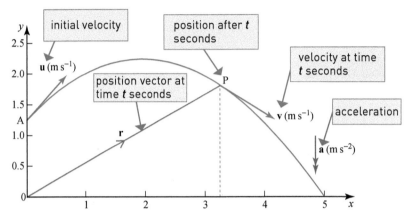

▲ Figure 1.4

The vector $\mathbf{r} = \overrightarrow{OP}$ is the position vector of the marble after a time t seconds and the vector $\mathbf{v}$ represents its velocity in $\mathrm{m\,s^{-1}}$ at that instant of time.

> Notice that the graph shows the trajectory of the marble. It is its path through space, not a position–time graph.

You can use equations for constant acceleration in vector form to describe the motion.

velocity

$$\mathbf{v} = \mathbf{u} + \mathbf{a}t$$

$$\mathbf{v} = \mathbf{u} + \mathbf{a}t$$

▲ Figure 1.5

> Always check whether or not the projectile starts at the origin. The change in position is the vector $\mathbf{r} - \mathbf{r}_0$. This is the equivalent of $s - s_0$ in one dimension.

displacement

$$\mathbf{r} - \mathbf{r}_0 = \mathbf{u}t + \tfrac{1}{2}\mathbf{a}t^2$$
$$\text{so } \mathbf{r} = \mathbf{r}_0 + \mathbf{u}t + \tfrac{1}{2}\mathbf{a}t^2$$

This is what the displacement would be without gravity.

initial displacement

This is the distance 'fallen' due to gravity:
$$\mathbf{a} = \begin{pmatrix} 0 \\ -g \end{pmatrix}$$

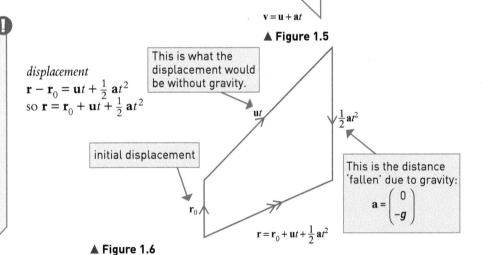

$$\mathbf{r} = \mathbf{r}_0 + \mathbf{u}t + \tfrac{1}{2}\mathbf{a}t^2$$

▲ Figure 1.6

When working with projectile problems, you can treat each direction separately or you can write them both together as vectors. The next example shows both methods.

Example 1.1

A ball is thrown horizontally at $5\,\mathrm{m\,s^{-1}}$ out of a window $4\,\mathrm{m}$ above the ground.

(i)   How long does it take to reach the ground?

(ii) How far from the building does it land?

(iii) What is its speed just before it lands and at what angle to the ground is it moving?

Solution

Figure 1.7 shows the path of the ball. It is important to decide at the outset where the origin and axes are. You may choose any axes that are suitable, but you must specify them carefully to avoid making mistakes. Here the origin is taken to be at ground level below the point of projection of the ball and upwards is positive. With these axes, the acceleration is $-g\,\mathrm{m\,s^{-2}}$.

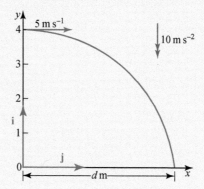

▲ **Figure 1.7**

Method 1: Resolving into components

(i)   *Position*: Using axes as shown and $s = s_0 + ut + \frac{1}{2}at^2$ in the two directions,

Horizontally: $x_0 = 0, u_x = 5, a_x = 0$
$$x = 5t \qquad \text{①}$$

Vertically: $y_0 = 4, u_y = 0, a_y = -10$
$$y = 4 - 5t^2 \qquad \text{②}$$

> s_0 is the initial position, so the displacement at time t is $s - s_0$.

The ball reaches the ground when $y = 0$. Substituting $y = 0$ in equation ② gives

$$0 = 4 - 5t^2$$
$$t^2 = \frac{4}{5}$$
$$t = 0.894\ldots$$

> Remember that 0.894... means 0.894 and subsequent figures. This number is kept in your calculator for use in subsequent working.

The ball hits the ground after $0.894\,\mathrm{s}$.

(ii) When the ball lands $x = d$ so, from equation ①,

$$d = 5t = 5 \times 0.894... = 4.47...$$

The ball lands 4.47 m from the building.

(iii) *Velocity*: Using $v = u + at$ in the two directions,

Horizontally $\qquad v_x = 5 + 0$
Vertically $\qquad v_y = 0 - 10t$

To find the speed and direction just before it lands:
The ball lands when $t = 0.894...$ so

$$v_x = 5 \text{ and } v_y = 10 \times 0.894... = -8.94....$$

The components of velocity are shown in Figure 1.8.
The speed of the ball is

$$\sqrt{5^2 + 8.94...^2} = 10.2 \, \text{m s}^{-1}$$

It hits the ground moving downwards at an angle α to the horizontal where

$$\tan \alpha = \frac{8.94...}{5}$$

$$\alpha = 60.8°$$

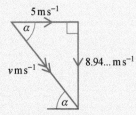

▲ **Figure 1.8**

Method 2: Using vectors

The initial position is $\mathbf{r}_0 = \begin{pmatrix} 0 \\ 4 \end{pmatrix}$ and the ball hits the ground when $\mathbf{r} = \begin{pmatrix} d \\ 0 \end{pmatrix}$.

The initial velocity, $\mathbf{u} = \begin{pmatrix} 5 \\ 0 \end{pmatrix}$ and the acceleration $\mathbf{a} = \begin{pmatrix} 0 \\ -10 \end{pmatrix}$.

Using $\qquad \mathbf{r} = \mathbf{r}_0 + \mathbf{u}t + \frac{1}{2}\mathbf{a}t^2$

$$\begin{pmatrix} d \\ 0 \end{pmatrix} = \begin{pmatrix} 0 \\ 4 \end{pmatrix} + \begin{pmatrix} 5 \\ 0 \end{pmatrix}t + \frac{1}{2}\begin{pmatrix} 0 \\ -10 \end{pmatrix}t^2$$

$$d = 5t \qquad\qquad ①$$

and $\qquad 0 = 4 - 5t^2 \qquad\qquad ②$

> These vectors could just as well have been written in terms of unit vectors $\mathbf{i}$ and $\mathbf{j}$ along the x and y directions.

(i) Equation ② gives $t = 0.894...$, so the ball hits the ground after 0.894 s.

(ii) Substituting this into ① gives $d = 4.47...$, so the ball lands 4.47 m from the building.

(iii) The speed and direction of motion are the magnitude and direction of the velocity of the ball. Using

$$\mathbf{v} = \mathbf{u} + \mathbf{a}t$$

$$\begin{pmatrix} v_x \\ v_y \end{pmatrix} = \begin{pmatrix} 5 \\ 0 \end{pmatrix} + \begin{pmatrix} 0 \\ -10 \end{pmatrix}t$$

So when $t = 0.894...$, $\begin{pmatrix} v_x \\ v_y \end{pmatrix} = \begin{pmatrix} 5 \\ -8.94... \end{pmatrix}$

You can find the speed and angle as in Method 1.

Notice that in both methods the time forms a link between the motions in the two directions. You can often find the time from one equation and then substitute it in another to find out more information.

Exercise 1B

In this exercise take upwards as positive.

1 In each case

  (a) draw a diagram showing the initial velocity and path

  (b) write the velocity after time t s in vector form

  (c) write the position after time t s in vector form.

  (i) Initial position $(0, 10\,\mathrm{m})$; initial velocity $4\,\mathrm{m\,s^{-1}}$ horizontally.

  (ii) Initial position $(0, 7\,\mathrm{m})$; initial velocity $10\,\mathrm{m\,s^{-1}}$ at $35°$ above the horizontal.

  (iii) Initial position $(0, 20\,\mathrm{m})$; initial velocity $10\,\mathrm{m\,s^{-1}}$ at $13°$ below the horizontal.

  (iv) Initial position O; initial velocity $\begin{pmatrix} 7 \\ 24 \end{pmatrix}\mathrm{m\,s^{-1}}$.

  (v) Initial position $(a, b)\,\mathrm{m}$; initial velocity $\begin{pmatrix} u_0 \\ v_0 \end{pmatrix}\mathrm{m\,s^{-1}}$.

2 In each case find

  (a) the time taken for the projectile to reach its highest point

  (b) the maximum height above the origin.

  (i) Initial position $(0, 15\,\mathrm{m})$; velocity $5\,\mathrm{m\,s^{-1}}$ horizontally and $14.7\,\mathrm{m\,s^{-1}}$ vertically.

  (ii) Initial position $(0, 10\,\mathrm{m})$; initial velocity $\begin{pmatrix} 5 \\ 3 \end{pmatrix}\mathrm{m\,s^{-1}}$.

3 Find the horizontal range for these projectiles, which start from the origin.

  (i) Initial velocity $\begin{pmatrix} 2 \\ 7 \end{pmatrix}\mathrm{m\,s^{-1}}$.

  (ii) Initial velocity $\begin{pmatrix} 7 \\ 2 \end{pmatrix}\mathrm{m\,s^{-1}}$.

  (iii) Sketch the paths of these two projectiles using the same axes.

CP 4 A projectile starts at $\begin{pmatrix} 0 \\ h \end{pmatrix}\mathrm{m}$ and is projected with initial velocity $\mathbf{u}\,\mathrm{m\,s^{-1}}$ at an angle θ to the horizontal. The particle experiences an acceleration $\begin{pmatrix} 0 \\ -g \end{pmatrix}\mathrm{m\,s^{-2}}$. Find the time taken for the projectile to hit the ground and its horizontal range.

Answers to exercises are available at www.hoddereducation.com/cambridgeextras

1.3 Further examples

| Example 1.2 | In this question, neglect air resistance. |

In an attempt to raise money for a charity, participants are sponsored to kick a ball over some vans. The vans are each 2.2 m high and 2 m wide and stand on horizontal ground. One participant kicks the ball at an initial speed of $24\,\text{m}\,\text{s}^{-1}$ inclined at 30° to the horizontal.

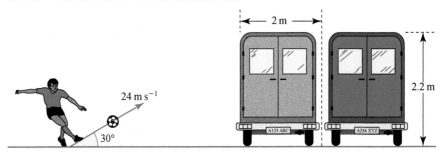

▲ Figure 1.9

(i) What are the initial values of the vertical and horizontal components of velocity?

(ii) Show that while in flight the vertical height y metres at time t seconds satisfies the equation $y = 12t - 5t^2$.
Calculate at what times the ball is at least 2.2 m above the ground.

The ball should pass over as many vans as possible.

(iii) Deduce that the ball should be placed about 4.2 m from the first van and find how many vans the ball will clear.

(iv) What is the greatest vertical distance between the ball and the top of the vans?

Solution

(i) *Initial velocity*
Horizontal component: $24\cos 30° = 20.7...\,\text{m}\,\text{s}^{-1}$
Vertical component: $24\sin 30° = 12\,\text{m}\,\text{s}^{-1}$

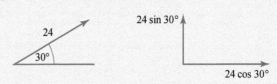

▲ Figure 1.10

(ii) *When the ball is above 2.2 m*
Using axes as shown and
$s = ut + \frac{1}{2}at^2$ vertically

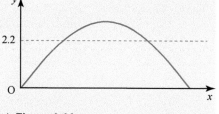

$$\Rightarrow \qquad y = 12t - 5t^2$$

The ball is 2.2 m above
the ground when $y = 2.2$, then

▲ **Figure 1.11**

$$2.2 = 12t - 5t^2$$
$$5t^2 - 12t + 2.2 = 0$$
$$25t^2 - 60t + 11 = 0$$
$$(5t - 1)(5t - 11) = 0$$
$$t = 0.2 \text{ or } 2.2$$

> $a = -10\,\text{m s}^{-2}$
> because the
> positive direction
> is upwards.

The ball is at least 2.2 m above the ground when $0.2 \leqslant t \leqslant 2.2$.

(iii) *How many vans?*
Horizontally, $s = ut + \frac{1}{2}at^2$ with

$a = 0$

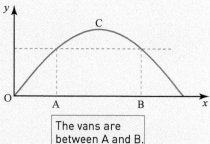

$$\Rightarrow \qquad x = 20.7...t$$

When $t = 0.2$, $\quad x = 4.15...$ (at A)

when $t = 2.2$, $\quad x = 45.7...$ (at B)

To clear as many vans as possible,
the ball should be placed about
4.2 m in front of the first van.

▲ **Figure 1.12**

> The vans are
> between A and B.

The distance between the first and last van cleared is
$\quad$ AB $= 45.7... - 4.15...\,\text{m} = 41.5...\,\text{m}$
$$\frac{41.5...}{2} = 20.7...$$

The maximum possible number of vans is 20.

(iv) *Maximum height*
At the top (C), vertical velocity $= 0$, so using $v = u + at$ vertically

$$\Rightarrow \quad 0 = 12 - 10t$$

$$t = 1.2$$

Substituting in $y = 12t - 5t^2$, maximum height is

$$12 \times 1.2 - 5 \times 1.2^2 = 7.2\,\text{m}$$

The greatest vertical distance between the ball and the top of the vans is

$$7.2 - 2.2 = 5\,\text{m}.$$

Answers to exercises are available at www.hoddereducation.com/cambridgeextras

Example 1.3

Sharon is diving into a swimming pool. During her flight she may be modelled as a particle. Her initial velocity is $1.8\,\text{m\,s}^{-1}$ at an angle of $30°$ above the horizontal and initial position $3.1\,\text{m}$ above the water. Air resistance may be neglected.

(i) Find the greatest height above the water that Sharon reaches during her dive.

(ii) Show that the time t, in seconds, that it takes Sharon to reach the water is given by $5t^2 - 0.9t - 3.1 = 0$ and solve the equation to find t. Explain the significance of the other root of the equation.

Just as Sharon is diving a small boy jumps into the swimming pool. He hits the water at a point in line with the diving board and $1.5\,\text{m}$ from its end.

(iii) Is there an accident?

Solution

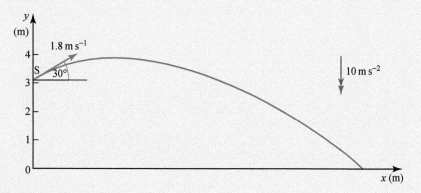

▲ **Figure 1.13**

Referring to the axes shown:

	Horizontal motion	Vertical motion
initial position	0	3.1
a	0	-10
u	$u_x = 1.8\cos 30° = 1.55\ldots$	$u_y = 1.8\sin 30° = 0.9$
v	$v_x = 1.55\ldots$ ①	$v_y = 0.9 - 10t$ ②
r	$x = 1.55\ldots t$ ③	$y = 3.1 + 0.9t - 5t^2$ ④

(i) At the top $v_y = 0$ $\quad 0 = 0.9 - 10t \implies t = 0.09$ $\qquad$ from ②

When $t = 0.09$

$\quad y = 3.1 + 0.9 \times 0.09 - 5 \times 0.09^2 = 3.14\ldots$ $\qquad$ from ④

Sharon's greatest height above the water is $3.14\,\text{m}$.

(ii) Sharon reaches the water when $y = 0$

$$0 = 3.1 + 0.9t - 5t^2 \qquad \text{from ④}$$

$$5t^2 - 0.9t - 3.1 = 0$$

$$t = \frac{0.9 \pm \sqrt{0.9^2 + 4 \times 5 \times 3.1}}{10}$$

$$t = -0.702\ldots \text{ or } 0.882\ldots$$

Sharon hits the water after $0.882\,\text{s}$. The negative value of t gives the point on the parabola at water level to the left of the point where Sharon dives.

(iii) At time t the horizontal distance from the diving board is

$$x = 1.55\ldots t \qquad \text{from ③}$$

When Sharon hits the water

$$x = 1.55\ldots \times 0.88\ldots = 1.37\ldots = 1.38$$

Assuming that the particles representing Sharon and the boy are located at their centres of mass, the difference of $12\,\text{cm}$ between $1.38\,\text{m}$ and $1.5\,\text{m}$ is not sufficient to prevent an accident.

Note

When the point at which Sharon dives is taken as the origin in this example, the initial position is $(0, 0)$ and $y = 0.9t - 5t^2$. In this case, Sharon hits the water when $y = -3.1$ m. This gives the same equation for t.

Example 1.4

A boy kicks a small ball from the floor of a gymnasium with an initial velocity of $15\,\text{m\,s}^{-1}$ inclined at an angle α to the horizontal. Air resistance may be neglected.

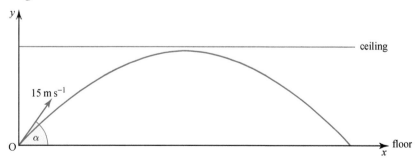

▲ Figure 1.14

(i) Write down expressions in terms of α for the vertical speed of the ball and vertical height of the ball after t seconds.

The ball just fails to touch the ceiling, which is $5\,\text{m}$ high. The highest point of the motion of the ball is reached after T seconds.

Answers to exercises are available at www.hoddereducation.com/cambridgeextras

(ii) Use one of your expressions to show that $3\sin\alpha = 2T$ and the other to form a second equation involving $\sin\alpha$ and T.

(iii) Eliminate $\sin\alpha$ from your two equations to show that T has a value of 1.

(iv) Find the horizontal range of the ball when it is kicked at $15\,\text{m s}^{-1}$ from the floor of the gymnasium so that it just misses the ceiling.

Solution

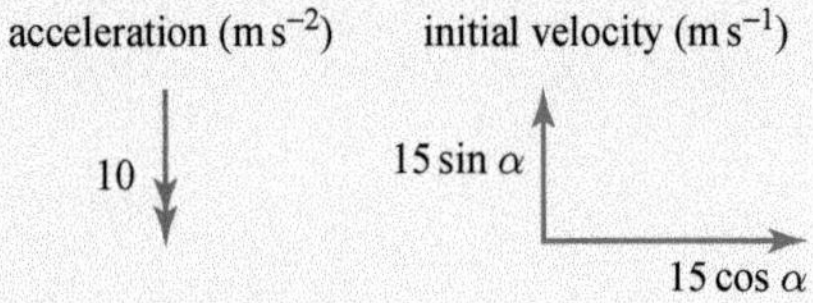

(i) *Vertical components*

speed $\quad v_y = 15\sin\alpha - 10t$ ①

height $\quad y = (15\sin\alpha)t - 5t^2$ ②

acceleration (m s^{-2})   initial velocity (m s^{-1})

$10\downarrow$ $15\sin\alpha$

$15\cos\alpha$

▲ **Figure 1.15**

(ii) *Time to highest point*

At the top $v_y = 0$ and $t = T$, so equation ① gives

$15\sin\alpha - 10T = 0$

$15\sin\alpha = 10T$

$3\sin\alpha = 2T$ ③

When $t = T, y = 5$, so from   ②

$5 = (15\sin\alpha)T - 5T^2$ ④

(iii) Substituting for $3\sin\alpha$ from ③ into ④ gives

$5 = 2\times 5T\times T - 5T^2$

$5 = 5T^2$

$T = 1$

(iv) *Range*

The path is symmetrical so the time of flight is $2T$ seconds.

Horizontally $a = 0$ and $u_x = 15\cos\alpha$

$\Rightarrow x = (15\cos\alpha)t$

The range is $15\cos\alpha \times 2T = 30\cos\alpha\,\text{m}$.

From ③ $3\sin\alpha = 2T = 2$

$\sin\alpha = \frac{2}{3}$

$\Rightarrow \cos\alpha = \sqrt{1 - \frac{4}{9}} = \frac{1}{3}\sqrt{5}$

The range is $30 \times \frac{1}{3}\sqrt{5} = 22.4\,\text{m}$.

> Two marbles start simultaneously from the same height. One (P) is dropped and the other (Q) is projected horizontally. Which reaches the ground first?

1 A ball is thrown from a point at ground level with velocity $20\,\text{m s}^{-1}$ at 30° to the horizontal. The ground is level and horizontal and you should ignore air resistance.

(i) Find the horizontal and vertical components of the ball's initial velocity.

(ii) Find the horizontal and vertical components of the ball's acceleration.

(iii) Find the horizontal distance travelled by the ball before its first bounce.

(iv) Find how long the ball takes to reach maximum height.

(v) Find the maximum height reached by the ball.

2 A golf ball is hit with a velocity of $45\,\text{m s}^{-1}$ at an elevation of 30°, along a level fairway. For the golf ball, find

(i) the greatest height reached

(ii) the time of flight

(iii) the distance travelled along the fairway.

3 Ying hits a golf ball with initial velocity $50\,\text{m s}^{-1}$ at 35° to the horizontal.

(i) Find the horizontal and vertical components of the ball's initial velocity.

(ii) Specify suitable axes and calculate the position of the ball at one second intervals for the first six seconds of its flight.

(iii) Draw a graph of the path of the ball (its trajectory) and use it to estimate

(a) the maximum height of the ball

(b) the horizontal distance the ball travels before bouncing.

(iv) Calculate the maximum height the ball reaches and the horizontal distance it travels before bouncing. Compare your answers with the estimates you found from the graph drawn in (iii).

(v) State the modelling assumptions you made in answering this question.

PS 4 Clare scoops a hockey ball off the ground, giving it an initial velocity of $19\,\text{m s}^{-1}$ at 25° to the horizontal.

(i) Find the horizontal and vertical components of the ball's initial velocity.

(ii) Find the time that elapses before the ball hits the ground.

(iii) Find the horizontal distance the ball travels before hitting the ground.

(iv) Find how long it takes for the ball to reach maximum height.

(v) Find the maximum height reached.

A member of the opposing team is standing 20 m away from Clare in the direction of the ball's flight. The opposing player can hold her hockey stick so that it reaches a maximum distance of 2.5 m above the ground.

(vi) How high above the ground is the ball when it passes the opposing player? Can she stop the ball with her hockey stick?

Answers to exercises are available at www.hoddereducation.com/cambridgeextras

PS 5 A footballer is standing 30 m in front of the goal. He kicks the ball towards the goal with velocity $18 \, \text{m s}^{-1}$ and angle 55° to the horizontal. The height of the goal's crossbar is 2.5 m. Air resistance and spin may be neglected.

(i) Find the horizontal and vertical components of the ball's initial velocity.

(ii) Find the time it takes for the ball to cross the goal-line.

(iii) Does the ball bounce in front of the goal, go straight into the goal or go over the crossbar?

In fact the goalkeeper is standing 5 m in front of the goal and will stop the ball if its height is less than 2.8 m when it reaches him.

(iv) Does the goalkeeper stop the ball? Justify your answer.

PS 6 A plane is flying at a speed of $300 \, \text{m s}^{-1}$ and maintaining an altitude of 10 000 m when a bolt becomes detached. Ignoring air resistance, find

(i) the time that the bolt takes to reach the ground

(ii) the horizontal distance between the point where the bolt leaves the plane and the point where it hits the ground

(iii) the speed of the bolt when it hits the ground

(iv) the angle to the horizontal at which the bolt hits the ground.

7 A particle P is projected with speed $26 \, \text{m s}^{-1}$ at an angle of 30° below the horizontal, from a point O which is 80 m above horizontal ground.

(i) Calculate the distance from O of the particle 2.3 s after projection.

(ii) Find the horizontal distance travelled by P before it reaches the ground.

(iii) Calculate the speed and direction of motion of P immediately before it reaches the ground.

Cambridge International AS & A Level Mathematics
9709 Paper 52 Q6 June 2011

8 A particle P is released from rest at a point A which is 7 m above horizontal ground. At the same instant that P is released a particle Q is projected from a point O on the ground. The horizontal distance of O from A is 24 m. Particle Q moves in the vertical plane containing O and A, with initial speed $50 \, \text{m s}^{-1}$ and initial direction making an angle θ above the horizontal, where $\tan \theta = \frac{7}{24}$ (see diagram). Show that the particles collide.

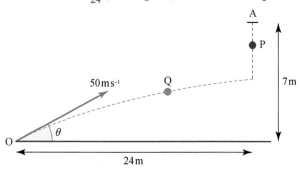

Cambridge International AS & A Level Mathematics
9709 Paper 52 Q3 November 2009

9   A particle P is projected with speed $V \, \text{m s}^{-1}$ at an angle of $60°$ above the horizontal from a point O on horizontal ground. P is moving at an angle of $45°$ above the horizontal at the instant $1.5 \, \text{s}$ after projection.

(i)   Find V.

(ii) Hence calculate the horizontal and vertical displacements of P from O at the instant $1.5 \, \text{s}$ after projection.

Cambridge International AS & A Level Mathematics
9709 Paper 53 Q2 June 2015

PS

10 To kick a goal in rugby you must kick the ball over the crossbar of the goal posts (height $3 \, \text{m}$), between the two uprights. Dafydd attempts a kick from a distance of $35 \, \text{m}$. The initial velocity of the ball is $20 \, \text{m s}^{-1}$ at $30°$ to the horizontal. The ball is aimed between the uprights and no spin is applied.

(i)   How long does it take for the ball to reach the goal posts?

(ii) Does the ball go over the crossbar? Justify your answer.

Later in the game Daffyd takes another kick from the same position and hits the crossbar.

(iii) Given that the initial velocity of the ball in this kick was also at $30°$ to the horizontal, find the initial speed.

PS

11 Reena is learning to serve in tennis. She hits the ball from a height of $2 \, \text{m}$. For her serve to be legal it must pass over the net, which is $12 \, \text{m}$ away from her and $0.91 \, \text{m}$ high, and it must land within $6.4 \, \text{m}$ of the net.

Use the following modelling assumptions to answer this question.

»   She hits the ball horizontally.

»   Air resistance may be ignored.

»   The ball may be treated as a particle.

»   The ball does not spin.

(i)   How long does the ball take to fall to the level of the top of the net?

(ii) How long does the ball take from being hit to first reaching the ground?

(iii) What is the lowest initial speed of the ball that allows it to pass over the net?

(iv) What is the greatest initial speed of the ball if it lands within $6.4 \, \text{m}$ of the net?

12 A stunt motorcycle rider attempts to jump over a gorge $50 \, \text{m}$ wide. He takes off from a ramp angled at $25°$ to the horizontal at a speed of $30 \, \text{m s}^{-1}$.

(i)   Assuming that air resistance is negligible, find out whether the rider crosses the gorge successfully. You must justify your answer.

The stuntman believes that, in any jump, the effect of air resistance reduces his horizontal range by 40%.

(ii) Allowing for air resistance, calculate his minimum safe take-off speed for this jump.

13 A ball is kicked from a point A on level ground and hits a wall at a point 4 m above the ground. The wall is at a distance of 32 m from A. Initially the velocity of the ball makes an angle of arctan $\frac{3}{4}$ with the ground. Find the initial speed of the ball and its speed when it hits the wall.

(M)

14 A catapult projects a small pellet at speed $20\,\mathrm{m\,s^{-1}}$ and can be directed at any angle to the horizontal.

    (i)   Find the range of the catapult when the angle of projection is

          (a) 30°   (b) 40°   (c) 45°   (d) 50°   (e) 60°.

    (ii)   Show algebraically that, when the angle of projection is α, the range is the same as it is when the angle of projection is $90 - \alpha$.

The catapult is angled with the intention that the pellet should hit a point on the ground 36 m away.

    (iii) Verify that one appropriate angle of projection would be 32.1° and write down another suitable angle.

In fact the angle of projection from the catapult is liable to error.

    (iv) Find the distance by which the pellet misses the target in each of the cases in (iii) when the angle of projection is subject to an error of ±0.5°. Which angle should you use for greater accuracy?

(PS)

15 A cricketer hits the ball from ground level. The ball leaves the ground at 30° to the horizontal and travels towards a fielder standing on the boundary 60 m away.

    (i)   Find the initial speed of the ball if it hits the ground for the first time at the fielder's feet.

    (ii)   Find the initial speed of the ball if it is at a height of 3.2 m (well outside the fielder's reach) when it passes over the fielder's head.

In fact the fielder is able to catch the ball without moving provided that its height, h m, when it reaches him satisfies the inequality $0.25 \leqslant h \leqslant 2.1$.

    (iii) Find a corresponding range of values of u, the initial speed of the ball, that allow the fielder to catch the ball.

(PS)

16 A horizontal tunnel has a height of 3 m. A ball is thrown inside the tunnel with an initial speed of $18\,\mathrm{m\,s^{-1}}$. What is the greatest horizontal distance that the ball can travel before it bounces for the first time?

1.4 The path of a projectile

Look at the equations

$$x = 20t$$

$$y = 6 + 30t - 5t^2$$

They represent the path of a projectile.

> What is the initial velocity of the projectile? What is its initial position? What value of g is assumed?

These equations give x and y in terms of a third variable, t. (They are called **parametric equations** and t is the **parameter**.)

You can find the **Cartesian equation** connecting x and y directly by eliminating t as follows.

$$x = 20t \Rightarrow t = \frac{x}{20}$$

So

$$y = 6 + 30t - 5t^2$$

can be written as

$$y = 6 + 30 \times \frac{x}{20} - 5 \times \left(\frac{x}{20}\right)^2$$

$$y = 6 + 1.5x - \frac{x^2}{80} \longleftarrow \boxed{\text{This is the Cartesian equation.}}$$

Exercise 1D

1 Find the Cartesian equation of the path of each of these projectiles by eliminating the parameter t.

   (i) $x = 4t$ $y = 5t^2$

   (ii) $x = 5t$ $y = 6 + 2t - 5t^2$

   (iii) $x = 2 - t$ $y = 3t - 5t^2$

   (iv) $x = 1 + 5t$ $y = 8 + 10t - 5t^2$

   (v) $x = ut$ $y = 2ut - \frac{1}{2}gt^2$

CP 2 A particle is projected with initial velocity $50\,\text{m s}^{-1}$ at an angle of $36.9°$ to the horizontal. The point of projection is taken to be the origin, with the x-axis horizontal and the y-axis vertical in the plane of the particle's motion.

   (i)   Show that at time t s, the height of the particle in metres is given by

$$y = 30t - 5t^2$$

        and write down the corresponding expression for x.

   (ii)   Eliminate t between your equations for x and y to show that

$$y = \frac{3x}{4} - \frac{x^2}{320}.$$

   (iii) Plot the graph of y against x, using a scale of $2\,\text{cm}$ for $10\,\text{m}$ along both axes.

   (iv) Mark on the graph drawn in (iii) the points corresponding to the position of the particle after $1, 2, 3, 4, \ldots$ seconds.

3 A golfer hits a ball with initial velocity $50\,\mathrm{m\,s^{-1}}$ at an angle α to the horizontal where $\sin\alpha = 0.6$.

(i) Find the equation of its trajectory, assuming that air resistance may be neglected.

The flight of the ball is recorded on film and its position vector, from the point where it was hit, is calculated. The results (to the nearest $0.5\,\mathrm{m}$) are as shown in the table.

Time (s)	0	1	2	3	4	5	6
Position (m)	$\begin{pmatrix}0\\0\end{pmatrix}$	$\begin{pmatrix}39.5\\24.5\end{pmatrix}$	$\begin{pmatrix}78\\39\end{pmatrix}$	$\begin{pmatrix}116.5\\44\end{pmatrix}$	$\begin{pmatrix}152\\39\end{pmatrix}$	$\begin{pmatrix}187.5\\24.5\end{pmatrix}$	$\begin{pmatrix}222\\0\end{pmatrix}$

(ii) On the same piece of graph paper draw the trajectory you found in part (i) and that which you found from analysing the film. Compare the two graphs and suggest a reason for any differences.

(iii) It is suggested that the horizontal component of the resistance to the motion of the golf ball is almost constant. Are the figures consistent with this?

CP **4** A particle is projected from a point O with initial velocity having components u_x and u_y along the horizontal and vertical directions, respectively.

(i) If (x, y) is a point on the trajectory of the projectile, show that
$$y(u_x)^2 - u_x u_y x + 5x^2 = 0.$$

(ii) Find the speed of projection and the elevation, if the particle passes through the points with coordinates $(2, 1)$ and $(10, 1)$.

1.5 General equations

The work done in this chapter can now be repeated for the general case, using algebra. Assume a particle is projected from the origin with speed u at an angle α to the horizontal and that the only force acting on the particle is the force due to gravity. The x- and y-axes are horizontal and vertical through the origin, O, in the plane of motion of the particle.

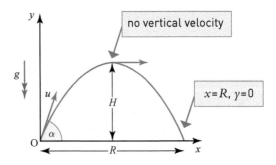

▲ **Figure 1.16**

The components of velocity and position

	Horizontal motion		Vertical motion	
Initial position	0		0	
a	0		$-g$	
u	$u_x = u\cos\alpha$		$u_y = u\sin\alpha$	
v	$v_x = u\cos\alpha$	①	$v_y = u\sin\alpha - gt$	②
r	$x = ut\cos\alpha$	③	$y = ut\sin\alpha - \frac{1}{2}gt^2$	④

> $ut\cos\alpha$ is preferable to $u\cos\alpha t$ because this could mean $u\cos(\alpha t)$, which is incorrect.

The maximum height

At its greatest height, the vertical component of velocity is zero.

From equation ②

$$u\sin\alpha - gt = 0$$

$$t = \frac{u\sin\alpha}{g}$$

Substitute in equation ④ to obtain the height of the projectile

$$y = u \times \frac{u\sin\alpha}{g} \times \sin\alpha - \frac{1}{2}g \times \frac{(u\sin\alpha)^2}{g^2}$$

$$= \frac{u^2\sin^2\alpha}{g} - \frac{u^2\sin^2\alpha}{2g}$$

The greatest height is

$$H = \frac{u^2\sin^2\alpha}{2g}$$

The time of flight

When the projectile hits the ground, $y = 0$.

From equation ④
$$y = ut\sin\alpha - \frac{1}{2}gt^2$$

$$0 = ut\sin\alpha - \frac{1}{2}gt^2$$

$$0 = t\left(u\sin\alpha - \frac{1}{2}gt\right)$$

> The solution $t = 0$ is at the start of the motion.

$$t = 0 \text{ or } t = \frac{2u\sin\alpha}{g}$$

The time of flight is $\qquad T = \dfrac{2u\sin\alpha}{g}$

The range

The range of the projectile is the value of x when $t = \dfrac{2u\sin\alpha}{g}$

From equation ④ : $\qquad x = ut\cos\alpha$

$$\Rightarrow \qquad R = u \times \frac{2u\sin\alpha}{g} \times \cos\alpha$$

$$R = \frac{2u^2\sin\alpha\cos\alpha}{g}$$

Answers to exercises are available at www.hoddereducation.com/cambridgeextras

It can be shown that $2\sin\alpha\cos\alpha = \sin 2\alpha$, so the range can be expressed as

$$R = \frac{u^2 \sin 2\alpha}{g}$$

The range is a maximum when $\sin 2\alpha = 1$, that is when $2\alpha = 90°$ or $\alpha = 45°$. The maximum possible horizontal range for projectiles with initial speed u is

$$R_{\text{max}} = \frac{u^2}{g}$$

The equation of the path

From equation ③ $\qquad t = \dfrac{x}{u\cos\alpha} = \dfrac{x}{u}\sec\alpha$

Substitute into equation ④ to give

$$y = u \times \frac{x}{u\cos\alpha} \times \sin\alpha - \frac{1}{2}g \times \frac{x^2}{u^2}\sec^2\alpha$$

$$y = x\frac{\sin\alpha}{\cos\alpha} - \frac{gx^2}{2u^2}\sec^2\alpha$$

So the equation of the trajectory is

$$y = x\tan - \frac{gx^2}{2u^2}\sec^2\alpha$$

Using the identity $\sec^2\alpha = 1 + \tan^2\alpha$ gives

$$y = x\tan - \frac{gx^2}{2u^2}(1 + \tan^2\alpha)$$

!

It is important that you understand the methods used to derive these formulae and don't rely on learning the results by heart. They are only true when the given assumptions apply and the variables are as defined in Figure 1.16 on page 20.

?

❯ What are the assumptions on which this work is based?

In this exercise, use the modelling assumptions that air resistance can be ignored and the ground is horizontal.

1   A projectile is launched from the origin with an initial velocity $30\,\mathrm{m\,s}^{-1}$ at an angle of $45°$ to the horizontal.

   (i)   Write down the coordinates of the position of the projectile after time t.

   (ii)   Show that the equation of the path is the parabola $y = x - \dfrac{1}{90}x^2$.

   (iii)   Find y when $x = 10$.

   (iv)   Find x when $y = 20$.

2   Jack throws a cricket ball with velocity $10\,\mathrm{m\,s}^{-1}$ at $14°$ above the horizontal. The ball leaves his hand $1.5\,\mathrm{m}$ above the origin.

   (i)   Show that the path of the ball is given by
   $$y = 1.5 + 0.25x - 0.053x^2.$$

   Jack is aiming at a stump $0.7\,\mathrm{m}$ high.

   (ii)   How far from the stump is he standing if the ball just hits the top?

3   While practising his tennis serve, Matthew hits the ball from a height of $2.5\,\mathrm{m}$ with a velocity of magnitude $25\,\mathrm{m\,s}^{-1}$ at an angle of $5°$ above the horizontal, as shown in the diagram.

   (i)   Show that while in flight
   $$y = 2.5 + 0.087x - 0.0081x^2.$$

   (ii)   Using the equation given in (i), find the horizontal distance from the serving point to the spot where the ball lands.

   (iii)   Determine whether the ball would clear the net, which is $1\,\mathrm{m}$ high and $12\,\mathrm{m}$ from the serving position in the horizontal direction. Justify your answer.

4   Ching is playing volleyball. She hits the ball with initial speed $u\,\mathrm{m\,s}^{-1}$ from a height of $1\mathrm{m}$ at an angle of $35°$ to the horizontal.

   (i)   Define a suitable origin and x-and y-axes and find the equation of the trajectory of the ball in terms of x, y and u. In your equation, give coefficients of x as decimals to 3 s.f.

   The rules of the game require the ball to pass over the net, which is at height $2\,\mathrm{m}$, and land inside the court on the other side, which is of length $5\,\mathrm{m}$ from the net to the baseline. Ching hits the ball perpendicular to the net and is $3\,\mathrm{m}$ from the net when she does so.

   (ii)   Using your equation from (i), find the minimum value of u for the ball to pass over the net.

   (iii)   Using your equation from (i), find the maximum value of u for the ball to land inside the court.

Answers to exercises are available at www.hoddereducation.com/cambridgeextras

5   The equation of the trajectory of a projectile that is projected from a point P is given by

$$y = 1 + 0.16x - 0.008x^2$$

where y is the height of the projectile above horizontal ground and x is the horizontal displacement of the projectile from P.
The projectile hits the ground at a point Q.

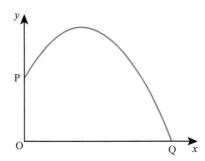

(i)   Write down the height of P and find the coordinates of Q.

(ii) Find the horizontal distance x from P of the highest point of the trajectory and show that this point is 1.8 m above ground.

(iii) Find the time taken for the projectile to fall from its highest point to the ground.

(iv) Find the horizontal component of the velocity. Deduce from this the time of flight for the projectile to travel from P to Q.

(v)   Calculate the speed of the projectile when it hits the ground.

6   A particle is projected from a point O on horizontal ground. The velocity of projection has magnitude $20\,\text{m}\,\text{s}^{-1}$ and direction upwards at an angle θ to the horizontal. The particle passes through the point which is 7 m above the ground and 16 m horizontally from O, and hits the ground at the point A.

(i)   Using the equation of the particle's trajectory and the identity $\sec^2\theta = 1 + \tan^2\theta$, show that the possible values of $\tan\theta$ are $\frac{3}{4}$ and $\frac{17}{4}$.

(ii) Find the distance OA for each of the two possible values of $\tan\theta$.

(iii) Sketch in the same diagram the two possible trajectories.

Cambridge International AS & A Level Mathematics
9709 Paper 51 Q5 June 2010

7   A particle P is projected with speed $25\,\text{m}\,\text{s}^{-1}$ at an angle of 45° above the horizontal from a point O on horizontal ground. At time t seconds after projection the horizontal and vertically upward displacements of P from O are $x\,\text{m}$ and $y\,\text{m}$, respectively.

(i)   Express x and y in terms of t and hence show that the equation of the path of P is $y = x - 0.016x^2$.

(ii) Calculate the horizontal distance between the two positions at which P is 2.4 m above the ground.

Cambridge International AS & A Level Mathematics
9709 Paper 53 Q3 November 2011

8   A small ball is thrown horizontally with speed 5 m s^{-1} from a point O on the roof of a building.

At time t s after projection, the horizontal and vertically downwards displacements of the ball from O are x m and y m, respectively.

(i)   Express x and y in terms of t, and hence show that the equation of the trajectory of the ball is $y = 0.2x^2$.

The ball strikes the horizontal ground which surrounds the building at a point A.

(ii) Given that OA $= 18$ m, calculate the value of x at A, and the speed of the ball immediately before it strikes the ground at A.

Cambridge International AS & A Level Mathematics
9709 Paper 53 Q5 June 2014

9   A particle is projected from a point O and passes through a point P on its trajectory when it is travelling horizontally. The coordinates of P are (16, 12). Find the angle of projection and the magnitude of the initial velocity.

PS   10   A golf ball is driven from the tee with speed $30\sqrt{2}$ m s^{-1} at an angle α to the horizontal.

(i)   Show that during its flight the horizontal and vertical displacements x and y of the ball from the tee satisfy the equation
$$y = x \tan\alpha - \frac{x^2}{360}(1 + \tan^2\alpha).$$

(ii) The golf ball just clears a tree 5 m high that is 150 m horizontally from the tee. Find the two possible values of $\tan\alpha$.

(iii) Use the discriminant of the quadratic equation in $\tan\alpha$ to find the greatest distance by which the golf ball can clear the tree and find the value of $\tan\alpha$ in this case.

(iv) The ball is aimed at the hole that is on the green immediately behind the tree. The hole is 160 m from the tee. What is the greatest height the tree could be if it is possible to hit a hole in one?

CP   11   A boy is firing small stones from a catapult at a target on the top of a wall.

The stones are projected from a point that is 5 m from the wall and 1 m above ground level. The target is on top of the 3 m high wall. The stones are projected at a speed of 10 m s^{-1} at an angle of θ to the horizontal.

(i)   The stone hits the target. Show that θ must satisfy the equation
$$5\tan^2\theta - 20\tan\theta + 13 = 0.$$

(ii) Find the two values of θ for which the stone hits the target.

CP | **12** A shot-putter projects a small shot from a point 2 m above the ground, which is horizontal. The speed of projection is $10\,\text{m}\,\text{s}^{-1}$ and the angle of projection is θ above the horizontal.

(i) Show that the time, in seconds, that elapses before the shot hits the ground is $\sqrt{\dfrac{1-c}{2}} + \sqrt{\dfrac{9-5c}{10}}$, where $c = \cos 2\theta$.

(ii) Find an expression for the range in terms of c and show that it is greatest when $c = \dfrac{1}{6}$.

(iii) Find the maximum range and compare it to the range found when $\theta = 45°$.

KEY POINTS

1 Modelling assumptions for projectile motion with acceleration due to gravity:

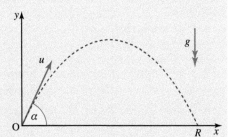

- a projectile is a particle
- it is not powered
- the air has no effect on its motion.

2 Projectile motion is usually considered in terms of horizontal and vertical components.

When the initial position is O

Angle of projection = α

Initial velocity, $\mathbf{u} = \begin{pmatrix} u \cos \alpha \\ u \sin \alpha \end{pmatrix}$

Acceleration, $\mathbf{g} = \begin{pmatrix} 0 \\ -g \end{pmatrix}$

At time t, velocity, $\mathbf{v} = \mathbf{u} + \mathbf{a}t$ $\begin{pmatrix} v_x \\ v_y \end{pmatrix} = \begin{pmatrix} u \cos \alpha \\ u \sin \alpha \end{pmatrix} + \begin{pmatrix} 0 \\ -g \end{pmatrix}t$

$$v_x = u \cos \alpha \qquad \text{①}$$
$$v_y = u \sin \alpha - gt \qquad \text{②}$$

Displacement, $\mathbf{r} = \mathbf{u}t + \frac{1}{2}\mathbf{a}t^2$ $\begin{pmatrix} x \\ y \end{pmatrix} = \begin{pmatrix} u \cos \alpha \\ u \sin \alpha \end{pmatrix}t + \frac{1}{2}\begin{pmatrix} 0 \\ -g \end{pmatrix}t^2$

$$x = ut \cos \alpha \qquad \text{③}$$
$$y = ut \sin \alpha - \frac{1}{2}gt^2$$

3   At a maximum height $v_y = 0$.

4   $y = 0$ when the projectile lands.

5   The time to hit the ground is twice the time to maximum height.

6   The equation of the path of a projectile is

$$y = x \tan \alpha - \frac{gx^2}{2u^2}(1 + \tan^2 \alpha)$$

7   When the point of projection is (x_0, y_0) rather than $(0, 0)$

$$\mathbf{r} = \mathbf{r}_0 + \mathbf{u}t + \tfrac{1}{2}\mathbf{a}t^2 \qquad \begin{pmatrix} x \\ y \end{pmatrix} = \begin{pmatrix} x_0 \\ y_0 \end{pmatrix} + \begin{pmatrix} u\cos\alpha \\ u\sin\alpha \end{pmatrix} t + \tfrac{1}{2}\begin{pmatrix} 0 \\ -g \end{pmatrix} t^2$$

LEARNING OUTCOMES

Now that you have finished this chapter, you should be able to

- model the motion of a projectile as a particle moving with constant acceleration
- use horizontal and vertical equations of motion to solve problems on the motion of a projectile
- find the magnitude and direction of the velocity at any given time or position
- find the range of the projectile on a horizontal plane
- find the greatest height reached by the projectile
- derive the Cartesian equation of the trajectory of the projectile
- use the equation of the trajectory in solving problems in which the initial speed or angle of projection may be unknown.

2 Moments of forces

2

> Give me a firm place to stand and I will move the Earth.
> *Archimedes*
> *(c. 287–c. 212BC)*

> › The photograph shows a swing bridge over a canal. It can be raised to allow barges and boats to pass. It is operated by hand, even though it is very heavy. How is this possible?

The bridge depends on the turning effects or **moments** of forces. To understand these you might find it helpful to look at a simpler situation.

Two children sit on a simple see-saw, made of a plank balanced on a fulcrum, as in Figure 2.1. Will the see-saw balance?

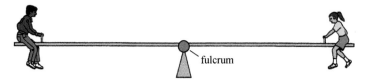
fulcrum

▲ **Figure 2.1**

If both children have the same mass and sit the same distance from the fulcrum, then you expect the see-saw to balance.

Now consider possible changes to this situation.

(i)   If one child is heavier than the other, you expect the heavier one to go down.

(ii) If one child moves nearer the centre, you expect that child to go up.

You can see that both the weights of the children and their distances from the fulcrum are important.

What about this case? One child has mass 35 kg and sits 1.6 m from the fulcrum and the other has mass 40 kg and sits on the opposite side 1.4 m from the fulcrum, as in Figure 2.2.

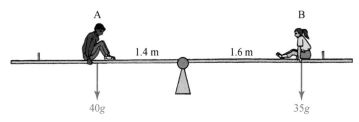

▲ **Figure 2.2**

Taking the products of their weights and their distances from the fulcrum gives

A: $40g \times 1.4 = 56g$
B: $35g \times 1.6 = 56g$

So you might expect the see-saw to balance, and this indeed is what would happen.

2.1 Rigid bodies

Until now the particle model has provided a reasonable basis for the analysis of the situations you have met. In examples like the see-saw, however, where turning is important, this model is inadequate because the forces do not all act through the same point.

In such cases you need the **rigid body model**, in which an object, or **body**, is recognised as having size and shape, but is assumed not to be deformed when forces act on it.

Suppose a tray is lying on a smooth table. Then, using one finger, you push the tray so that the force acts parallel to one edge and through the centre of mass, as in Figure 2.3.

▲ **Figure 2.3**

The particle model is adequate here: the tray travels in a straight line in the direction of the applied force.

If you push the tray equally hard with two fingers, as in Figure 2.4, symmetrically either side of the centre of mass, the particle model is still adequate.

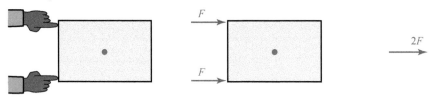

▲ **Figure 2.4**

However, if the two forces are not equal or are not symmetrically placed or, as in Figure 2.5, are in different directions, the particle model cannot be used.

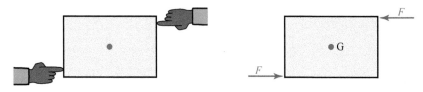

▲ **Figure 2.5**

The resultant force is now zero, since the individual forces are equal in magnitude but opposite in direction. What happens to the tray? Experience tells you that it starts to rotate about G. How quickly it starts to rotate depends, among other things, on the magnitude of the forces and the width of the tray. The rigid body model allows you to analyse the situation.

2.2 Moments

In the example of the see-saw you looked at the product of each force and its distance from a fixed point. This product is called the moment of the force about the point.

The see-saw balances because the moments of the forces on either side of the fulcrum have the same magnitude and act in opposite directions. One would tend to make the see-saw turn clockwise, the other anticlockwise. By contrast, the moments about G of the forces on the tray in the last situation do not balance. They both tend to turn it anticlockwise, so rotation occurs.

Conventions, notation and units

The moment of a force F about a point O is defined by

$$\text{moment} = Fd$$

where d is the perpendicular distance from the point O to the line of action of the force (Figure 2.6).

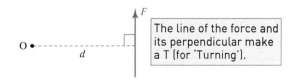

The line of the force and its perpendicular make a T (for 'Turning').

▲ **Figure 2.6**

In two dimensions, the sense of a moment is described as either positive (anticlockwise) or negative (clockwise), as shown in Figure 2.7.

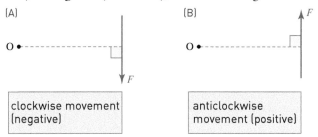

(A)

clockwise movement (negative)

(B)

anticlockwise movement (positive)

▲ **Figure 2.7**

If you imagine putting a pin at O and pushing along the line of F, your page would turn clockwise for (A) and anticlockwise for (B).

In the S.I. system the unit for moment is the newton metre (Nm), because a moment is the product of a force, the unit of which is the Newton, and distance, the unit of which is the metre.

Remember that moments are always taken about a point and you must always specify what that point is. A force acting through the point will have no moment about that point because in that case $d = 0$.

> Figure 2.8 shows two tools for undoing wheel nuts on a car. Discuss the advantages and disadvantages of each.

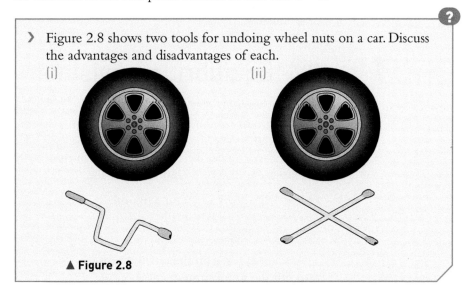

(i)   (ii)

▲ **Figure 2.8**

When using the spider wrench (the tool with two 'arms'), you apply equal and opposite forces either side of the nut. These produce moments in the same direction. One advantage of this method is that there is no resultant force and hence no tendency for the nut to snap off.

2.3 Couples

Whenever two forces of the same magnitude act in opposite directions along different lines, they have a zero resultant force, but do have a turning effect. In fact the moment will be Fd about any point, where d is the perpendicular distance between the forces. This is demonstrated in Figure 2.9.

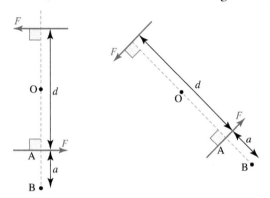

▲ **Figure 2.9**

In each of these situations:

Moment about O $\qquad$ $F\dfrac{d}{2} + F\dfrac{d}{2} = Fd$ ◄──────── Anticlockwise is positive.

Moment about A $\qquad\qquad 0 + Fd = Fd$

Moment about B $\qquad -aF + (a + d)\,F = Fd$

Any set of forces like these, with a zero resultant but a non-zero total moment, is known as a **couple**. The effect of a couple on a rigid body is to cause rotation.

2.4 Equilibrium revisited

In earlier work, an object was said to be in equilibrium if the resultant force on the object is zero. This definition is adequate provided all the forces act through the same point on the object. However, you are now concerned with forces acting at different points and, in this situation, even if the forces balance there may be a resultant moment.

Figure 2.10 shows a tray, on a smooth surface, being pushed equally hard at opposite corners.

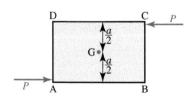

▲ **Figure 2.10**

The resultant force on the tray is clearly zero, but the resultant moment about its centre point, G, is

$$P \times \frac{a}{2} + P \times \frac{a}{2} = Pa.$$

The tray will start to rotate about its centre and so it is clearly not in equilibrium.

So the mathematical definition of equilibrium now needs to be tightened, to include moments. For an object to remain in equilibrium (either at rest or moving at constant velocity) when a system of forces is applied, both the resultant force and the total moment must be zero.

To check that an object is in equilibrium under the action of a system of forces, you need to check two things:

1 that the resultant force is zero;

2 that the resultant moment about any point is zero. (You only need to check one point.)

Example 2.1

Two children are playing with a door. Kerry tries to open it by pulling on the handle with a force of 50 N at right angles to the plane of the door, at a distance 0.8 m from the hinges. Peter pushes at a point 0.6 m from the hinges, also at right angles to the door and with sufficient force just to stop Kerry opening it.

(i) What is the moment of Kerry's force about the hinges?

(ii) With what force does Peter push?

(iii) Describe the resultant force on the hinges.

▲ **Figure 2.11**

Solution

Looking down from above, the line of the hinges becomes a point, H. The door opens clockwise. Anticlockwise is taken to be positive.

(i)

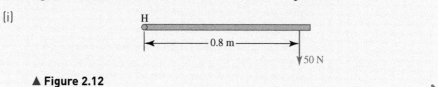

▲ **Figure 2.12**

Answers to exercises are available at www.hoddereducation.com/cambridgeextras

Kerry's moment about H $= -50 \times 0.8$
$$= -40\,\text{Nm}$$

The moment of Kerry's force about the hinges is $-40\,\text{Nm}$.
(Note that it is a clockwise moment and so negative.)

(ii)

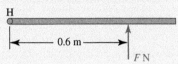

H

|← 0.6 m →|

$F\,\text{N}$

▲ **Figure 2.13**

Peter's moment about H $= + F \times 0.6$

Since the door is in equilibrium, the total moment on it must
be zero, so

$$F \times 0.6 - 40 = 0$$
$$F = \frac{40}{0.6}$$
$$= 66.7$$

Peter pushes with a force of $66.7\,\text{N}$.

(iii) Since the door is in equilibrium the overall resultant force on it
must be zero.

All the forces are at right angles to the door, as shown in Figure 2.14.

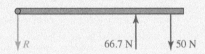

R $66.7\,\text{N}$ $50\,\text{N}$

▲ **Figure 2.14**

Resolve perpendicular to door:

$$R + 50 = 66.6\ldots$$

$$R = 16.7$$

The total reaction at the hinges is a force of $16.7\,\text{N}$ in the same
direction as Kerry is pulling in.

Note

The reaction force at a hinge may act in any
direction, according to the forces elsewhere
in the system. A hinge can be visualised in
cross section, as shown in Figure 2.15. If the
hinge is well oiled, and the friction between
the inner and outer parts is negligible, the
hinge cannot exert any moment. In this
situation the door is said to be 'freely hinged'.

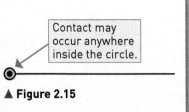

Contact may
occur anywhere
inside the circle.

▲ **Figure 2.15**

| Example 2.2 | The diagram shows a man of weight 600 N standing on a footbridge that consists of a uniform wooden plank just over 2 m long, of weight 200 N. Find the reaction forces exerted on each end of the plank. |

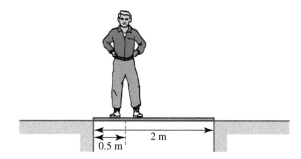

2 m
0.5 m

▲ **Figure 2.16**

Solution

The diagram shows the forces acting on the plank.

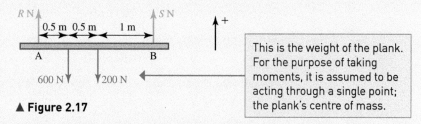

R N S N

0.5 m   0.5 m   1 m

A   B

600 N   200 N

This is the weight of the plank. For the purpose of taking moments, it is assumed to be acting through a single point; the plank's centre of mass.

▲ **Figure 2.17**

For equilibrium both the resultant force and the total moment must be zero.

Forces

$$R + S - 800 = 0 \qquad ①$$

> All forces act vertically.

Moments

Taking moments about the point A gives

$$(\curvearrowright) \quad R \times 0 - 600 \times 0.5 - 200 \times 1 + S \times 2 = 0 \qquad ②$$

> You could take moments about B instead – or any other point.

From equation ② $S = 250$ and so equation ① gives $R = 550$.

The reaction forces are 250 N at A and 550 N at B.

Notes

1 You cannot solve this problem without taking moments.

2 You can take moments about any point and can, for example, show that by taking moments about B you get the same answer.

3 The whole weight of the plank is being considered to act at its centre.

4 When a force acts through the point about which moments are being taken, its moment about that point is zero.

Answers to exercises are available at www.hoddereducation.com/cambridgeextras

Levers

A lever can be used to lift or move a heavy object by using a relatively small force. Levers depend on moments for their action.

Two common lever configurations are shown below. In both cases a load W is being lifted by an applied force F, using a lever of length l. The calculations assume equilibrium.

Type 1

The fulcrum is within the lever, as shown in Figure 2.18.

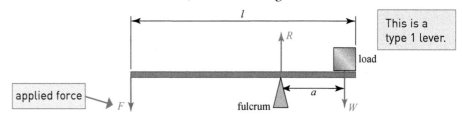

▲ **Figure 2.18**

Taking moments about the fulcrum:

$$(\curvearrowleft) \quad F \times (l - a) - W \times a = 0$$

$$F = W \times \frac{a}{l - a}$$

Provided that the fulcrum is nearer the end with the load, the applied force is less than the load.

Examples of type 1 levers are seesaws, crowbars and scissors.

Type 2

The fulcrum is at one end of the lever, as shown in Figure 2.19.

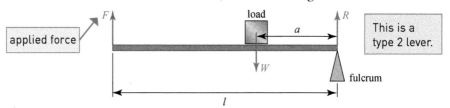

▲ **Figure 2.19**

Taking moments about the fulcrum:

$$(\curvearrowright) \quad F \times l - W \times a = 0$$

$$F = W \times \frac{a}{l}$$

Examples of type 2 levers include wheel barrows, nutcrackers and bottle openers.

Since a is much smaller than l, the applied force F is much smaller than the load W.

These examples also indicate how to find a single force equivalent to two parallel forces. The force equivalent to F and W should be equal and opposite to R and have the same line of action.

Example 2.3

Describe the single force equivalent to P and Q in each of these cases. In each case state its magnitude and line of action.

(i)

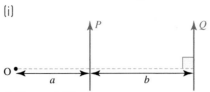

(ii)

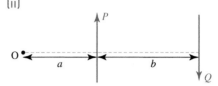

▲ **Figure 2.20**

Solution

(i)

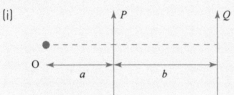

▲ **Figure 2.21**

The resultant of the two forces P and Q is a force of magnitude $P + Q$ pointing upwards.

The total moment of the forces P and Q about O is $P \times a + Q \times (a + b).$

Replacing P and Q by a single force $(P + Q)$ requires placing it at a distance x from O, such that $x(P + Q) = Pa + Q(a + b),$

leading to $x = \dfrac{Pa + Q(a + b)}{P + Q}$

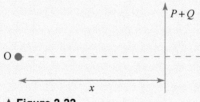

▲ **Figure 2.22**

The single force equivalent to P and Q is shown in Figure 2.22.

(ii)

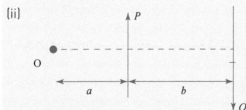

▲ **Figure 2.23**

The resultant of the two forces P and Q $(P > Q)$ is the force of magnitude $(P - Q)$ pointing upwards. The sign of the resultant force would be different for $P < Q$.

The total moment of the forces P and Q about O is $P \times a - Q \times (a + b).$

Replacing P and Q by a single force $(P - Q)$ requires placing it at a distance y from O such that $(P - Q)y = Pa - Q(a + b)$

leading to $y = \dfrac{Pa - Q(a + b)}{P - Q}.$

Notice that if $y < 0$, it means that the force $(P - Q)$ is to the left of O.

➡

Answers to exercises are available at www.hoddereducation.com/cambridgeextras

The single force equivalent to P and Q is the force $(P - Q)$ pointing upwards and distant $\dfrac{Pa - Q(a+b)}{P-Q}$ from O.

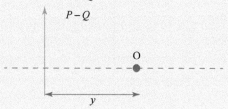

▲ **Figure 2.24**

> ⟩ How do you use moments to open a screw-top jar?
> ⟩ Why is it an advantage to press hard when the lid is stiff?

Exercise 2A

1 In each of the situations shown below, find the moment of the force about the point and state whether it is positive (anticlockwise) or negative (clockwise).

(i)

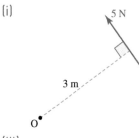

(ii)

(iii)

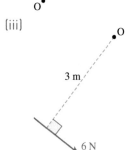

(iv)

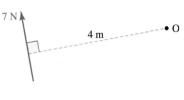

2 The situations below involve several forces acting on each object. For each one, find the total moment.

(i)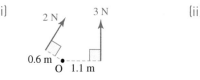

(ii)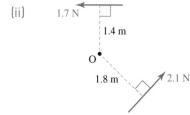

(iii)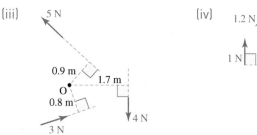

(iv)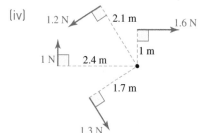

3 A uniform horizontal bar of mass 5 kg has length 30 cm and rests on two vertical supports, 10 cm and 22 cm from its left-hand end. Find the magnitude of the reaction force at each of the supports.

4 The diagram shows a motorcycle of mass 250 kg and its rider, whose mass is 80 kg. The centre of mass of the motorcycle lies on a vertical line midway between its wheels. When the rider is on the motorcycle, his centre of mass is 1 m behind the front wheel.

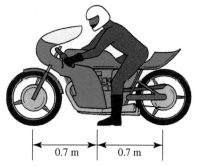

Find the vertical reaction forces acting through the front and rear wheels when

(i) the rider is not on the motorcycle

(ii) the rider is on the motorcycle.

Answers to exercises are available at www.hoddereducation.com/cambridgeextras

5 Find the reaction forces X and Y acting on the hi-fi shelf shown below. The shelf itself has weight 25 N and its centre of mass is midway between A and D.

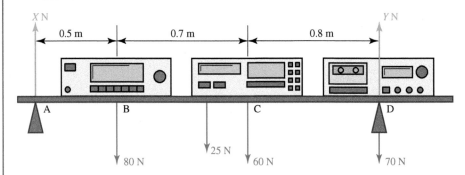

PS **6** Karen and Jane are trying to find the positions of their centres of mass. They place a uniform board of mass 8 kg symmetrically on two bathroom scales whose centres are 2 m apart. When Karen lies flat on the board, Jane notes that scale A reads 37 kg and scale B reads 26 kg.

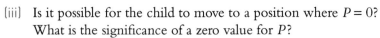

(i) Draw a diagram showing the forces acting on the board and calculate Karen's mass.

(ii) How far from the centre of scale A is her centre of mass?

M **7** The diagram shows two people, an adult and a child, sitting on a uniform bench of mass 40 kg. Their positions are as shown. The mass of the child is 50 kg and the mass of the adult is 85 kg.

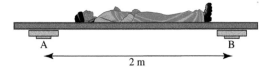

(i) Find the reaction forces, P and Q (in N), from the ground on the two supports of the bench.

(ii) The child now moves to the midpoint of the bench. What are the new values of P and Q?

(iii) Is it possible for the child to move to a position where $P = 0$? What is the significance of a zero value for P?

(iv) What happens if the child leaves the bench?

8 The diagram shows a diving board that some children have made. It consists of a uniform plank of mass 20 kg and length 3 m, with 1 m of its length projecting out over a pool. They have put a boulder of mass 25 kg above a support on the end over the land and there is a second support at the water's edge.

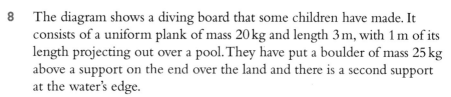
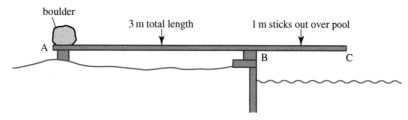

(i) Find the forces at the two supports when nobody is using the diving board.

(ii) A child of mass 50 kg is standing on the end of the diving board over the pool. What are the forces at the two supports?

(iii) Some older children arrive and take over the diving board. One of these is a heavy boy of mass 90 kg. What is the reaction at A if the board begins to tip over?

(iv) How far can the boy in part (iii) walk from B before the board tips over?

9 A lorry of mass 5000 kg is driven across a metal beam bridge of mass 20 tonnes. The bridge is a roadway of length 10 m that is supported at both ends.

(i) Find expressions for the reaction forces at each end of the bridge in terms of the distance x in metres travelled by the lorry from the start of the bridge.

(ii) From what point of the lorry is the distance x measured?

Two identical lorries cross the bridge at the same speed, starting at the same instant, from opposite directions.

(iii) How do the reaction forces of the supports on the bridge vary as the lorries cross the bridge?

10 A non-uniform rod AB of length 20 cm rests horizontally on two supports that are positioned at C and D, where AC = BD = 4 cm. The greatest mass that can be hung from A without disturbing equilibrium is 8 grams, and the greatest mass that can be hung from B is 10 grams. Find the mass of the rod and the distance of its centre of mass from A.

11 A uniform plank of mass 80 kg is 12 m long. The plank is laid on the ground near a quayside so that 5 m of the plank sticks out over the side of a quay. What is the minimum load that must be placed on the end of the plank that lies on the ground so that a woman of mass 45 kg can walk to the other end of the plank without tipping into the water?

12 A simple suspension bridge across a narrow river consists of a uniform beam, 4 m long and of mass 60 kg, supported by vertical cables attached at a distance 0.75 m from each end of the beam.

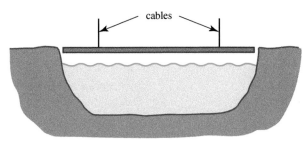

cables

(i) Find the tension in each cable when a boy of mass 50 kg stands 1 m from the end of the bridge.

(ii) Can a couple walking hand-in-hand cross the bridge safely, without it tipping, if their combined mass is 115 kg? Justify your answer.

(iii) What is the mass of a person standing on the end of the bridge when the tension in one cable is four times that in the other cable?

13 Find the magnitude, direction and line of action of the resultant of this system of forces.

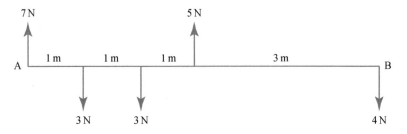

14 The diagram shows a stone slab AB of mass 1 tonne resting on two supports, C and D. The stone is uniform and has length 3 m. The supports are at distances 1.2 m from A and 0.5 m from B, as shown in the diagram.

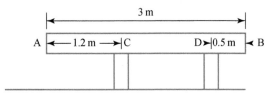

(i) Find the reaction forces at the two supports.

Local residents are worried that the arrangement is unsafe since their children play on the stone.

(ii) How many children, each of mass 50 kg, would need to stand at A in order for the stone to be on the point of tipping about C?

The stone's owner decides to move the support at C to a point nearer to A. To take the weight of the slab while doing this, he sets up the lever system shown in the diagram. The distance XF is 1.25 m and FY is 0.25 m. The end Y of the lever is immediately under point A on the stone slab. The lever can be modelled as a light rod.

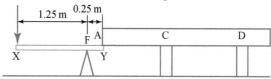

(iii) What downward force applied at X would reduce the reaction force at C to zero (and so allow the support to be moved)?

15 Four seamen are using a light capstan to pull in their ship's anchor at a steady rate. One of them is shown in the diagram. The anchor cable is wound around the capstan's drum, which has diameter 1 m. The spokes on which the men are pushing each project 1.5 m from the *centre* of the capstan. Each man is pushing with a force of 300 N, horizontally and at right angles to his spoke. You may assume that each man's force is applied at the end of his spoke.

The anchor cable is taut; it passes over a frictionless pulley and then makes an angle of 20° with the horizontal.

Note:
Only one of the four seamen is shown.

(i) Find the tension in the cable.

The mass of the ship is 2000 tonnes.

(ii) Find the acceleration of the ship, assuming that the ship moves only in the horizontal plane and that no other horizontal forces act on the ship.

In fact the acceleration of the ship is $0.0015\,\text{m s}^{-2}$. Part of the difference can be explained by friction with the capstan, resulting in a resisting moment of 150 Nm, the rest by the force of resistance, R N, to the ship passing through the water.

(iii) Find the value of R.

Answers to exercises are available at www.hoddereducation.com/cambridgeextras

2

Set up the apparatus shown in Figure 2.25 and experiment with two or more weights in different positions.

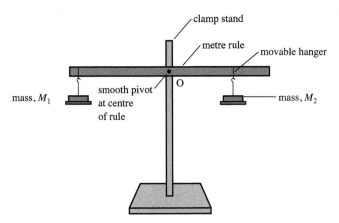

▲ **Figure 2.25**

Record your results in a table showing weights, distances from O and moments about O.

Two masses are suspended from the rule in such a way that the rule balances in a horizontal position. What happens when the rule is then moved to an inclined position and released?

Now attach a pulley, as in Figure 2.26. Start with equal weights and measure d and l. Then try different weights and pulley positions.

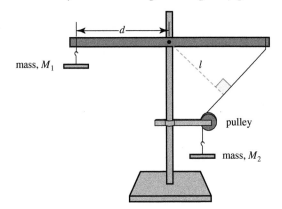

▲ **Figure 2.26**

The moment of a force that acts at an angle

From the experiment you will have seen that the moment of a force about the pivot depends on the **perpendicular distance** from the pivot to the line of the force.

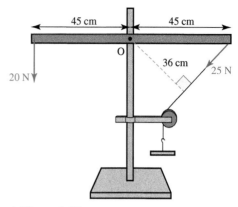

In Figure 2.27, where the system remains at rest, the moment about O of the 20 N force is $20 \times 0.45 = 9 \, \text{Nm}$.

The moment about O of the 25 N force is $-25 \times 0.36 = -9 \, \text{Nm}$.

▲ **Figure 2.27**

The system is in equilibrium even though unequal forces act at equal distances from the pivot.

The magnitude of the moment of the force F about O in Figure 2.28 is given by

$$F \times l = Fd \sin \alpha$$

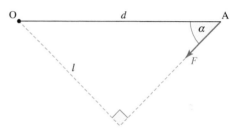

▲ **Figure 2.28**

Alternatively, the moment can be found by noting that the force F can be resolved into components $F \cos \alpha$ parallel to AO and $F \sin \alpha$ perpendicular to AO, both acting through A (Figure 2.29). The moment of each component can be found and then these moments can be summed to give the total moment. The moment of the component along AO is zero because it acts through O. The magnitude of the moment of the perpendicular component is $F \sin \alpha \times d$ so the total moment is $Fd \sin \alpha$, as expected.

▲ **Figure 2.29**

Example 2.4

A force of 40 N is exerted on a rod as shown. Find the moment of the force about the point marked O.

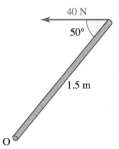

▲ Figure 2.30

Solution

In order to calculate the moment, the perpendicular distance between O and the line of action of the force must be found. This is shown on Figure 2.31.

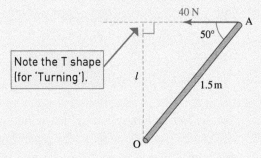

Note the T shape (for 'Turning').

▲ Figure 2.31

Here $l = 1.5 \times \sin 50°$.

So the moment about O is

$$F \times l = 40 \times (1.5 \times \sin 50°)$$
$$= 46.0\,\text{Nm}.$$

Alternatively, you can resolve the 40 N force into components, as in Figure 2.32.

The component of the force parallel to AO is 40 cos 50° N. The component perpendicular to AO is 40 sin 50° (or 40 cos 40°) N.

So the moment about O is

$$40 \sin 50° \times 1.5 = 60 \sin 50°$$
$$= 46.0\,\text{Nm as before.}$$

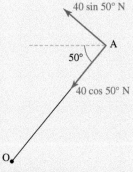

▲ Figure 2.32

Example 2.5

A sign is attached to a light rod of length 1 m which is freely hinged to the wall and supported in a vertical plane by a light string, as in Figure 2.33. The sign is assumed to be a uniform rectangle of mass 10 kg. The angle of the string to the horizontal is 25°.

▲ Figure 2.33

(i) Find the tension in the string.
(ii) Find the magnitude and direction of the reaction force of the hinge on the sign.

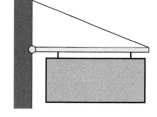

Solution

(i) Figure 2.34 shows the forces acting on the rod, where R_H and R_V are the magnitudes of the horizontal and vertical components of the reaction R on the rod at the wall.

▲ Figure 2.34

Taking moments about O:

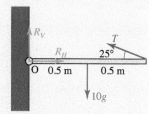

$$0 \times R_V + 0 \times R_H - 10g \times 0.5 + T \sin 25° \times 1 = 0$$
$$\Rightarrow \quad T \sin 25° = 5g$$
$$T = 118$$

The tension is 118 N.

(ii) You can resolve to find the reaction at the wall.

Horizontally: $R_H = T \cos 25°$

$\Rightarrow \quad R_H = 107$

Vertically: $R_V + T \sin 25° = 10g$

$\Rightarrow \quad R_V = 10g - 5g = 50$

$$R = \sqrt{107^2 + 50^2}$$
$$= 118$$
$$\tan \theta = \frac{50}{107}$$

$\theta = 25°$ (to the nearest degree)

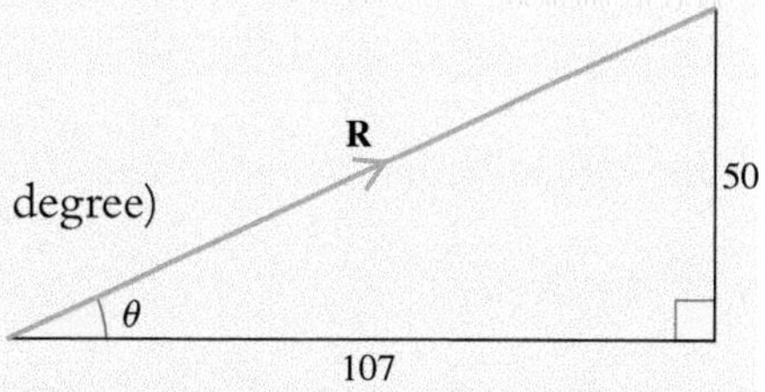

▲ Figure 2.35

The reaction at the hinge has magnitude 118 N and acts at 25° above the horizontal.

› Is it by chance that R and T have the same magnitude and act at the same angle to the horizontal?

Answers to exercises are available at www.hoddereducation.com/cambridgeextras

Example 2.6

A uniform ladder is standing on rough ground and leaning against a smooth wall at an angle of 60° to the ground. The ladder has length 4 m and mass 15 kg. Find the normal reaction forces at the wall and ground and the friction force at the ground.

Solution

Figure 2.36 shows the forces acting on the ladder. The forces are in newtons.

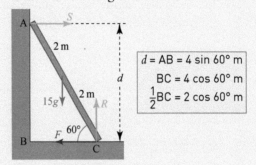

$d = AB = 4 \sin 60°$ m
$BC = 4 \cos 60°$ m
$\frac{1}{2}BC = 2 \cos 60°$ m

▲ **Figure 2.36**

Figure 2.36 shows that there are three unknown forces, S, R and F, so we need three equations from which to find them. If the ladder remains at rest (in equilibrium) then the resultant force is zero and the resultant moment is zero. These two conditions provide the three necessary equations.

Equilibrium of horizontal components: $\qquad S - F = 0 \qquad$ ①

Equilibrium of vertical components: $\qquad R - 15g = 0 \qquad$ ②

Moments about the foot of the ladder:

$R \times 0 + F \times 0 + 15g \times 2 \cos 60° - S \times 4 \sin 60° = 0$

$\Rightarrow \qquad\qquad 150 - 4S \sin 60° = 0 \qquad$ ③

$\Rightarrow \qquad\qquad\qquad\qquad\qquad\qquad S = \dfrac{150}{4 \sin 60°} = 43.3$

From ① $\qquad\qquad\qquad\qquad\qquad\qquad F = S = 43.3$

From ② $\qquad\qquad\qquad\qquad\qquad\qquad R = 150$

The force at the wall is 43.3 N.

The forces at the ground are 43.3 N horizontally and 150 N vertically.

1 Find the moment about O of each of the forces illustrated below.

(i)

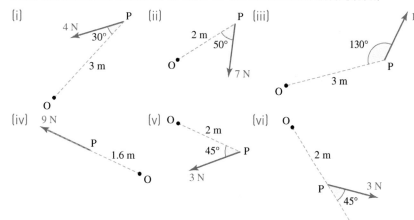

(ii)

(iii)

(iv)

(v)

(vi)

2 The diagram shows three children pushing a playground roundabout. Hannah and David want it to go one way but Rabina wants it to go the other way. Who wins? Justify your answer.

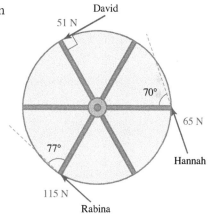

3 The operating regulations for a small crane specify that when the jib is at an angle of 25° above the horizontal, the maximum safe load for the crane is 5000 kg. Assuming that this maximum load is determined by the maximum moment that the pivot can support, what is the maximum safe load when the angle between the jib and the horizontal is

(i)   40°

(ii)   an angle θ?

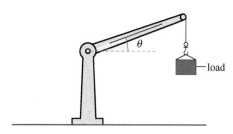

4 In each of these diagrams, a uniform beam of mass 5 kg and length 4 m, freely hinged at one end, A, is in equilibrium. Find the magnitude of the force T in each case.

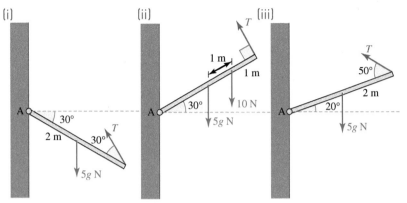

(i)   (ii)   (iii)

5 The diagram shows a uniform rectangular sign ABCD, 3 m × 2 m, of weight 20 N. It is freely hinged at A and supported by the string CE, which makes an angle of 30° with the horizontal. The tension in the string is T (in N).

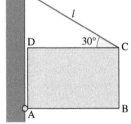

(i)   Resolve the tension T into horizontal and vertical components.

(ii) Hence show that the moment of the tension in the string about A is given by

$$2T \cos 30° + 3T \sin 30°.$$

(iii) Write down the moment of the sign's weight about A.

(iv) Hence show that $T = 9.28$ N.

(v)   Hence find the horizontal and vertical components of the reaction on the sign at the hinge, A.

You can also find the moment of the tension in the string about A as $d \times T$, where d is the length of AF as shown in the diagram.

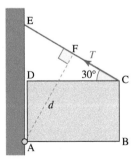

(vi) Find

(a)   the angle ACD   (b)   the length d.

(vii) Show that you get the same value for T when it is calculated in this way.

6 The diagram shows a simple crane. The weight of the jib (AB) may be ignored. The crane is in equilibrium in the position shown.

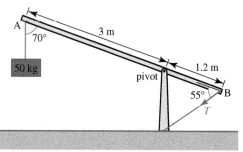

(i)   By taking moments about the pivot, find the magnitude of the tension T (in N).

(ii)   Find the reaction of the pivot on the jib in the form of components parallel and perpendicular to the jib.

(iii) Show that the total moment about the end A of the forces acting on the jib is zero.

(iv)   What would happen if

    (a)   the rope holding the 50 kg mass snapped?

    (b)   the rope with tension T snapped?

7 A uniform plank, AB, of mass 50 kg and length 6 m is in equilibrium leaning against a smooth wall at an angle of 60° to the horizontal. The lower end, A, is on rough horizontal ground.

(i)   Draw a diagram showing all the forces acting on the plank.

(ii)   Write down the total moment about A of all the forces acting on the plank.

(iii) Find the normal reaction of the wall on the plank at point B.

(iv)   Find the frictional force on the foot of the plank. What can you deduce about the coefficient of friction between the ground and the plank?

(v)   Show that the total moment about B of all the forces acting on the plank is zero.

8 A uniform ladder of mass 20 kg and length $2l$ rests in equilibrium with its upper end against a smooth vertical wall and its lower end on a rough horizontal floor. The coefficient of friction between the ladder and the floor is μ. The normal reaction at the wall is S, the frictional force at the ground is F and the normal reaction at the ground is R. The ladder makes an angle α with the horizontal.

(i)   Draw a diagram showing the forces acting on the ladder.

For each of the cases where   (a) $\alpha = 60°$ and   (b) $\alpha = 45°$

(ii)   find the magnitudes of S, F and R.

(iii) find the least possible value of μ.

Answers to exercises are available at www.hoddereducation.com/cambridgeextras

9 The diagram shows a car's handbrake. The force F is exerted by the hand in operating the brake, and this creates a tension T in the brake cable. The handbrake is freely pivoted at point B and is assumed to be light.

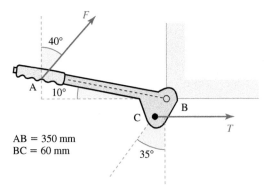

AB = 350 mm
BC = 60 mm

(i) Draw a diagram showing all the forces acting on the handbrake.

(ii) What is the required magnitude of force F if the tension in the brake cable is to be 1000 N?

(iii) A child applies the handbrake with a force of 10 N. What is the tension in the brake cable?

PS 10 The diagram shows four tugs manoeuvring a ship. A and C are pushing it, B and D are pulling it.

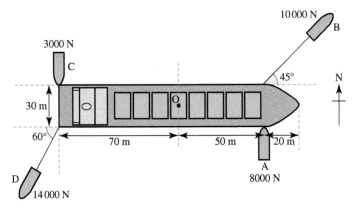

(i) Show that the resultant force on the ship is less than 100 N.

(ii) Find the overall turning moment on the ship about its centre point, O.

A breeze starts to blow from the south, causing a total force of 2000 N to act uniformly along the length of the ship, at right angles to it.

(iii) Assuming B and D continue to apply the same forces to the ship, how can tugs A and C counteract the sideways force on the ship by altering the forces with which they are pushing, while maintaining the same overall moment about the centre of the ship?

11 Chun is cleaning windows. Her ladder is uniform and stands on rough ground at an angle of 60° to the horizontal and with the top end resting on the edge of a smooth windowsill. The ladder has mass 12 kg and length 2.8 m and Chun has mass 56 kg.

(i)   Draw a diagram to show the forces on the ladder when nobody is standing on it. Show that the reaction at the sill is then $3g$ N.

(ii) Find the friction and normal reaction forces at the foot of the ladder.

Chun needs to be sure that the ladder will not slip, however high she climbs.

(iii) Find the least possible value of μ for the ladder to be safe at 60° to the horizontal.

(iv) The value of μ is in fact 0.4. How far up the ladder can Chun stand before it begins to slip?

12 A uniform rod AB of weight 16 N is freely hinged at A to a fixed point. A force of magnitude 4 N acting perpendicular to the rod is applied at B (see diagram).

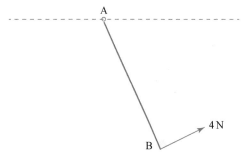

Given that the rod is in equilibrium

(i)   calculate the angle the rod makes with the horizontal

(ii) find the magnitude and direction of the force exerted on the rod at A.

Cambridge International AS & A Level Mathematics
9709 Paper 52 Q1 June 2011

CP 13 A uniform rod AB of length 6 m and weight 2000 N is hung from a point O by two light wires, each of length 5 m, attached to each end of the rod. A weight of 500 N is placed at a point C, 2 m from B. The tension in wire AO is T_1 and that in wire BO is T_2. The rod rests in equilibrium at an angle θ to the horizontal. The point X is directly below O and M is the midpoint of the rod.

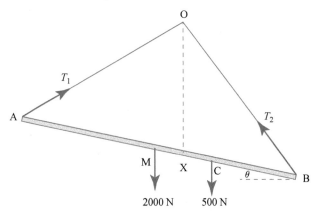

(i) By taking moments about O, find the distances MX and XC.

(ii) Find the angle θ.

(iii) By taking moments about each end of the rod, show that the ratio of the tensions in the wires is $T_1 : T_2 = 7 : 8$ and use this to find T_1 and T_2.

CP 14 The diagram shows a uniform ladder AB of mass m and length $2l$ resting in equilibrium with its upper end A against a smooth vertical wall and its lower end B on a smooth inclined plane. The inclined plane makes an angle θ with the horizontal and the ladder makes an angle ϕ with the wall. What is the relationship between θ and ϕ?

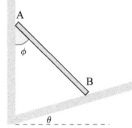

15 A uniform ladder of length 8 m and weight 180 N rests against a smooth, vertical wall and stands on a rough, horizontal surface. A woman of weight 720 N stands on the ladder so that her weight acts at a distance x m from its lower end, as shown in the diagram.

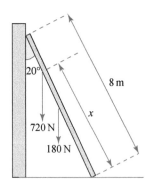

The system is in equilibrium with the ladder at 20° to the vertical.

(i) Show that the frictional force between the ladder and the horizontal surface is F N, where
$F = 90(1+x)\tan 20°$.

(ii) Deduce that F increases as x increases and hence find the values of the coefficient of friction between the ladder and the surface for which the woman can stand anywhere on the ladder without it slipping.

KEY POINTS

1 The moment of a force F about a point O is given by the product Fd, where d is the perpendicular distance from O to the line of action of the force.

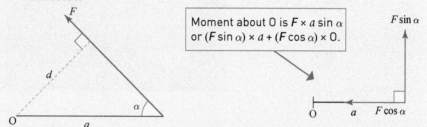

Moment about O is $F \times a \sin \alpha$ or $(F \sin \alpha) \times a + (F \cos \alpha) \times 0$.

2 The S.I. unit for moment is the newton metre (Nm).

3 Anticlockwise moments are usually called positive, clockwise negative.

4 If a body is in equilibrium, the sum of the moments of the forces acting on it, about any point, is zero.

LEARNING OUTCOMES

Now that you have finished this chapter, you should be able to
- calculate the moment of a force acting about a fixed point O by finding the product of the force with the perpendicular distance from O to the line of action of the force. Alternatively, you can first resolve the force into components and then you need only consider the component of the force that does not go through O
- find the resultant of a set of parallel forces
- understand how different types of lever work
- understand the meaning of the word 'couple'
- understand that an object is in equilibrium if the resultant of all the applied forces acting on it is zero and the sum of their moments about any point is also zero.

Answers to exercises are available at www.hoddereducation.com/cambridgeextras

3 Centre of mass

› Figure 3.1, which is drawn to scale, shows a mobile suspended from the point P. The horizontal rods and the strings are light but the geometrically shaped pieces are made of uniform heavy card. Does the mobile balance? If it does, what can you say about the position of its centre of mass?

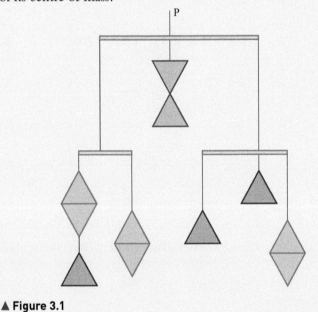

▲ Figure 3.1

In this chapter, you will consider the concept of centre of mass in the context of two general models.

» *The particle model*
The centre of mass is the single point at which the whole mass of the body may be taken to be situated.

» *The rigid body model*
The centre of mass is the balance point of a body with size and shape.

3.1 Centre of mass of a one-dimensional body

The following examples show how to calculate the position of the centre of mass of a body.

Example 3.1

An object consists of three point masses 8 kg, 5 kg and 4 kg attached to a rigid light rod, as shown in Figure 3.2.

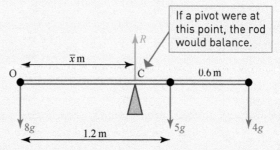

8 kg   1.2 m   5 kg   0.6 m   4 kg
O

▲ **Figure 3.2**

Calculate the distance of the centre of mass of the object from end O. (Ignore the mass of the rod.)

Solution

Suppose the centre of mass C is $\bar{x}$ m from O.

If a pivot were at this point, the rod would balance.

$\bar{x}$ m

O   C   0.6 m

8g   1.2 m   5g   4g

▲ **Figure 3.3**

For equilibrium $R = 8g + 5g + 4g = 17g$

Taking moments of the forces about O gives:

Total clockwise moment $\quad = (8g \times 0) + (5g \times 1.2) + (4g \times 1.8)$
$\qquad\qquad\qquad\qquad\quad = 13.2g$ Nm

Total anticlockwise moment $= R\bar{x}$
$\qquad\qquad\qquad\qquad\quad = 17g\bar{x}$ Nm

Answers to exercises are available at www.hoddereducation.com/cambridgeextras

The overall moment must be zero for the rod to be in balance, so

$$17g\bar{x} - 13.2g = 0$$
$$\Rightarrow \qquad 17\bar{x} = 13.2$$
$$\Rightarrow \qquad \bar{x} = \frac{13.2}{17} = 0.776$$

The centre of mass is $0.776\,\text{m}$ from the end O of the rod.

 Note

Although g was included in the calculation, it cancelled out. The answer depends only on the masses and their distances from the origin and not on the value of g.

Definition

This example can be generalised to give a method for finding the position of a centre of mass. Consider a set of n point masses $m_1, m_2, ..., m_n$ attached to a rigid light rod (whose mass is neglected) at positions $x_1, x_2, ..., x_n$ from one end O. The situation is shown in Figure 3.4.

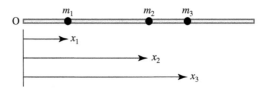

▲ **Figure 3.4**

The position, $\bar{x}$, of the centre of mass relative to O is defined by the equation:

moment of whole mass at centre of mass = sum of moments of individual masses

$$(m_1 + m_2 + ...m_n)\bar{x} = m_1x_1 + m_2x_2 + ...m_nx_n$$

The symbol Σ (sigma) means 'the sum of'.

or

$$M\bar{x} = \sum_{i=1}^{n} m_i x_i$$

where M is the total mass (or Σm_i).

| Example 3.2 |

A uniform rod of length $2\,\text{m}$ has mass $5\,\text{kg}$. Masses of $4\,\text{kg}$ and $6\,\text{kg}$ are fixed at each end of the rod. Find the centre of mass of the rod.

Solution

Since the rod is uniform, it can be treated as having a point mass at its centre. Figure 3.5 illustrates this situation.

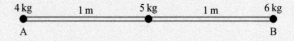

▲ **Figure 3.5**

Taking the end A as origin,

$$M\bar{x} = \sum m_i\, x_i$$
$$(4 + 5 + 6)\bar{x} = 4 \times 0 + 5 \times 1 + 6 \times 2$$
$$15\bar{x} = 17$$
$$\bar{x} = \frac{17}{15} = 1\frac{2}{15} = 1.13$$

So the centre of mass is $1.13\,\text{m}$ from the $4\,\text{kg}$ point mass.

> **?**
>
> ❯ Check that the rod in Example 3.2 would balance about a pivot $1\frac{2}{15}\text{m}$ from A.

Example 3.3

A rod AB of mass $1.1\,\text{kg}$ and length $1.2\,\text{m}$ has its centre of mass $0.48\,\text{m}$ from the end A. What mass should be attached to the end B to ensure that the centre of mass is at the midpoint of the rod?

Solution

Let the extra mass be $m\,\text{kg}$, as shown in Figure 3.6.

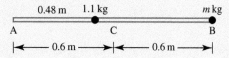

▲ **Figure 3.6**

Method 1

Refer to the midpoint, C, as origin, so $\bar{x} = 0$. Then

$$(1.1 + m) \times 0 = 1.1 \times (-0.12) + m \times 0.6$$
$$\Rightarrow \qquad 0.6m = 1.1 \times 0.12$$
$$\Rightarrow \qquad m = 0.22.$$

> The 1.1 mass has negative x referred to C.

A mass of 220 grams should be attached to B.

Method 2

Refer to the end, A, as origin, so $\bar{x} = 0.6$. Then

$$(1.1 + m) \times 0.6 = 1.1 \times 0.48 + m \times 1.2$$
$$\Rightarrow \qquad 0.66 + 0.6\,m = 0.528 + 1.2m$$
$$\Rightarrow \qquad 0.132 = 0.6m$$
$$m = 0.22 \text{ as before.}$$

Composite bodies

The position of the centre of mass of a composite body such as a cricket bat, tennis racquet or golf club is important to sportspeople, who like to feel its balance. If the body is symmetric then the centre of mass will lie on the axis of symmetry. The next example shows how to model a composite body as a system of point masses so that the methods of the previous section can be used to find the centre of mass.

Answers to exercises are available at www.hoddereducation.com/cambridgeextras

Example 3.4

A squash racquet of mass 200 g and total length 70 cm consists of a handle of mass 150 g, whose centre of mass is 20 cm from the end, and a frame of mass 50 g, whose centre of mass is 55 cm from the end.

Find the distance of the centre of mass from the end of the handle.

Solution

Figure 3.7 shows the squash racquet and its dimensions.

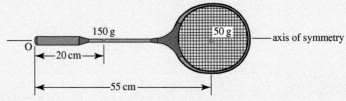

▲ Figure 3.7

The centre of mass lies on the axis of symmetry. Model the handle as a point mass of 0.15 kg at a distance 0.2 m from O and the frame as a point mass of 0.05 kg at a distance 0.55 m from the end O, as shown in Figure 3.8.

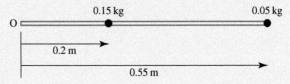

▲ Figure 3.8

The distance, $\bar{x}$, of the centre of mass from O is given by

$$(0.15 + 0.05)\,\bar{x} = (0.15 \times 0.2) + (0.05 \times 0.55)$$
$$\bar{x} = 0.288$$

The centre of mass of the squash racquet is 28.8 cm from the end of the handle.

Exercise 3A

1   The diagrams show point masses attached to rigid light rods. In each case, calculate the position of the centre of mass relative to the point O.

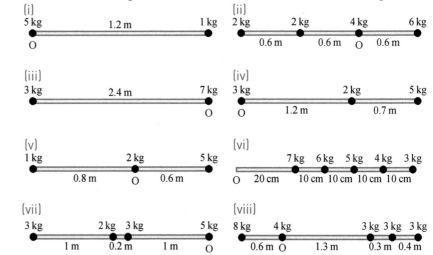

2 A see-saw consists of a uniform plank, 4 m long, of mass 10 kg. Calculate the position of the centre of mass when two children, of masses 20 kg and 25 kg, sit, one on each end.

3 A weightlifter's bar in a competition has mass 10 kg and length 1 m. By mistake, 50 kg is placed on one end and 60 kg on the other end. How far is the centre of mass of the bar from the centre of the bar itself?

4 The masses of the Earth and the Moon are 5.98×10^{24} kg and 7.38×10^{22} kg, and the distance between their centres is 3.84×10^{5} km. How far from the centre of the Earth is the centre of mass of the Earth–Moon system?

5 A crossing warden carries a sign that consists of a uniform rod of length 1.5 m and mass 1 kg, on top of which is a circular disc of radius 0.25 m and mass 0.2 kg. Find the distance of the centre of mass from the free end of the stick.

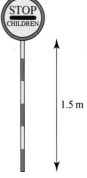

1.5 m

6 A rod has length 2 m and mass 3 kg. The centre of mass should be in the middle but, due to a fault in the manufacturing process, it is not. This error is corrected by placing a 200 g mass 5 cm from the centre of the rod. Where is the centre of mass of the rod itself?

7 A child's toy consists of four uniform discs, all made out of the same material. They each have thickness 2 cm and their radii are 6 cm, 5 cm, 4 cm and 3 cm. They are placed symmetrically on top of each other to form a tower. How high is the centre of mass of the tower?

8 A standard lamp consists of a uniform heavy metal base of thickness 4 cm, attached to which is a uniform metal rod of length 1.75 m and mass 0.25 kg.

What is the minimum mass for the base if the centre of mass of the lamp is no more than 12 cm from the ground?

9 A uniform scaffold pole of length 5 m has brackets bolted to it as shown in the diagram. The mass of each bracket is 1 kg.

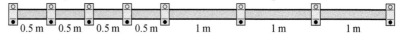

0.5 m   0.5 m   0.5 m   0.5 m   1 m   1 m   1 m

The centre of mass is 2.44 m from the left-hand end.
What is the mass of the pole?

10 An object of mass m_1 is placed at one end of a light rod of length l. An object of mass m_2 is placed at the other end. Find the distance of the centre of mass from m_1.

11 The diagram illustrates a mobile tower crane. It consists of the main vertical section (mass M tonnes), housing the engine, winding gear and controls, and the boom. The centre of mass of the main section is on its centre line. The boom, which has negligible mass, supports the load (L tonnes) and the counterweight (C tonnes). The main section stands on supports at P and Q, distance $2d$ m apart. The counterweight is held at a fixed distance a m from the centre line of the main section and the load at a variable distance l m.

In order for the crane to remain standing, the horizontal position of the centre of mass must remain between P and Q.

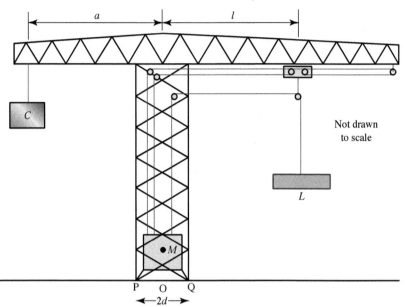

(i) In the case when $C = 3$, $M = 10$, $L = 7$, $a = 8$, $d = 2$ and $l = 13$, find the horizontal position of the centre of mass and say what happens to the crane.

(ii) Show that for these values of C, M, a, d and l the crane will not fall over when it has no load, and find the maximum safe load that it can carry.

(iii) Formulate two inequalities in terms of C, M, L, a, d and l that must hold if the crane is to be safe, loaded or unloaded.

(iv) Find, in terms of M, a, d and l, the maximum load that the crane can carry.

3.2 Centre of mass of two- and three-dimensional bodies

The techniques developed for using moments to find the centre of mass can be extended into two and three dimensions.

If a two-dimensional body consists of a set of n point masses $m_1, m_2, ..., m_n$ located at positions $(x_1, y_1), (x_2, y_2), ..., (x_n, y_n)$, as in Figure 3.9, then the position of the centre of mass of the body $(\bar{x}, \bar{y})$ is given by

$$M\bar{x} = \sum m_i x_i \quad \text{and} \quad M\bar{y} = \sum m_i y_i$$

where $M (= \sum m_i)$ is the total mass of the body.

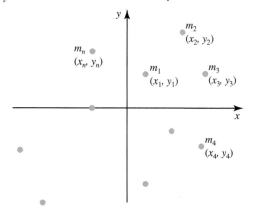

▲ Figure 3.9

In three dimensions, the z coordinates are also included. To find $\bar{z}$ use

$$M\bar{z} = \sum m_i z_i$$

The centre of mass of any composite body in two or three dimensions can be found by replacing each component by a point mass at its centre of mass.

| Example 3.5 | Joanna makes herself a pendant in the shape of a letter J, made up of rectangular shapes as shown in Figure 3.10. |

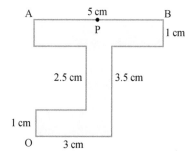

▲ Figure 3.10

(i) Find the position of the centre of mass of the pendant.

(ii) Find the angle that AB makes with the horizontal if she hangs the pendant from a point, P, in the middle of AB.

She wishes to hang the pendant so that AB is horizontal.

(iii) How far along AB should she place the ring through which the suspending chain will pass?

Answers to exercises are available at www.hoddereducation.com/cambridgeextras

Solution

(i)   The first step is to split the pendant into three rectangles.

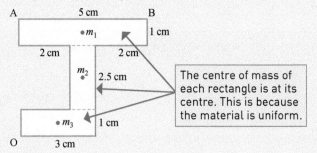

▲ **Figure 3.11**

You can model the pendant as three point masses m_1, m_2 and m_3, which are proportional to the areas of the rectangular shapes. Since the areas are $5\,cm^2$, $2.5\,cm^2$ and $3\,cm^2$, the masses, in suitable units, are 5, 2.5 and 3, and the total mass is $5 + 2.5 + 3 = 10.5$ (in the same units).

The table below gives the masses and positions of m_1, m_2 and m_3.

Mass		m_1	m_2	m_3	M
Mass units		5	2.5	3	10.5
Position of	**x**	2.5	2.5	1.5	$\bar{x}$
centre of mass	**y**	4	2.25	0.5	$\bar{y}$

Now it is possible to find $\bar{x}$:

$$M\bar{x} = \sum m_i\, x_i$$
$$10.5\bar{x} = 5 \times 2.5 + 2.5 \times 2.5 + 3 \times 1.5$$
$$\bar{x} = \frac{23.25}{10.5} = 2.21\,cm$$

Similarly for $\bar{y}$:

$$M\bar{y} = \sum m_i\, y_i$$
$$10.5\bar{y} = 5 \times 4 + 2.5 \times 2.25 + 3 \times 0.5$$
$$\bar{y} = \frac{27.125}{10.5} = 2.58\,cm$$

The centre of mass, G, is at (2.21, 2.58).

(ii) When the pendant is suspended from P, the centre of mass, G, is vertically below P, as shown in Figure 3.12 (opposite).

The pendant hangs like the first diagram, but you might find it easier to draw your own diagram like the second.

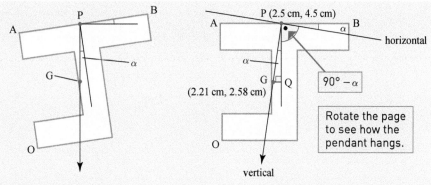

▲ Figure 3.12

$$GQ = 2.5 - 2.21 = 0.29$$
$$PQ = 4.5 - 2.58 = 1.92$$
$$\therefore \quad \tan \alpha = \frac{0.29}{1.92} \Rightarrow \alpha = 8.6°$$

Instead of rounding to three significant figures, it is usual to round angles to one decimal place.

AB makes an angle of 8.6° with the horizontal.

(iii) For AB to be horizontal the point of suspension must be directly above the centre of mass, and so it is 2.21 cm from A.

Example 3.6

Find the centre of mass of a body consisting of a square plate of mass 3 kg and side length 2 m, with small objects of mass 1 kg, 2 kg, 4 kg and 5 kg at the corners of the square.

Solution

Figure 3.13 shows the square plate, with the origin taken at the corner at which the 1 kg mass is located. The mass of the plate is represented by a 3 kg point mass at its centre.

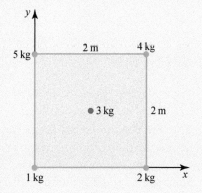

▲ Figure 3.13

In this example the total mass M (in kilograms) is $1 + 2 + 4 + 5 + 3 = 15$.

The two formulae for $\bar{x}$ and $\bar{y}$ can be combined into one using column vector notation:

$$\begin{pmatrix} M\bar{x} \\ M\bar{y} \end{pmatrix} = \begin{pmatrix} \Sigma m_i x_i \\ \Sigma m_i y_i \end{pmatrix}$$

which is equivalent to

$$M \begin{pmatrix} \bar{x} \\ \bar{y} \end{pmatrix} = \Sigma m_i \begin{pmatrix} x_i \\ y_i \end{pmatrix}$$

Substituting the known values for M and m_i and x_i and y_i:

$$15 \begin{pmatrix} \bar{x} \\ \bar{y} \end{pmatrix} = 1 \begin{pmatrix} 0 \\ 0 \end{pmatrix} + 2 \begin{pmatrix} 2 \\ 0 \end{pmatrix} + 4 \begin{pmatrix} 2 \\ 2 \end{pmatrix} + 5 \begin{pmatrix} 0 \\ 2 \end{pmatrix} + 3 \begin{pmatrix} 1 \\ 1 \end{pmatrix}$$

$$15 \begin{pmatrix} \bar{x} \\ \bar{y} \end{pmatrix} = \begin{pmatrix} 15 \\ 21 \end{pmatrix}$$

$$\begin{pmatrix} \bar{x} \\ \bar{y} \end{pmatrix} = \begin{pmatrix} 1 \\ 1.4 \end{pmatrix}$$

The centre of mass is at the point $(1, 1.4)$.

Example 3.7

A metal disc of radius $15\,\text{cm}$ has a hole of radius $5\,\text{cm}$ cut from it, as shown in Figure 3.14. Find the centre of mass of the disc.

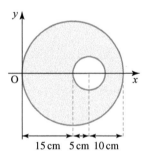

▲ **Figure 3.14**

15 cm   5 cm   10 cm

Solution

The disc is symmetrical about the central horizontal (x) axis. The centre of mass must therefore be on this axis. Think of the original uncut disc as a composite body made up of the final body and a disc to fit into the hole. Since the material is uniform, the mass of each part is proportional to its area.

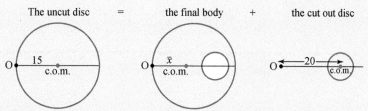

The uncut disc = the final body + the cut out disc

▲ **Figure 3.15**

	Uncut disc	Final body	Cut out disc
Area	$15^2\pi = 225\pi$	$15^2\pi - 5^2\pi = 200\pi$	$5^2\pi = 25\pi$
Distance from O to centre of mass	15 cm	$\bar{x}$ cm	20 cm

Taking moments about O:

$$225\pi \times 15 = 200\pi \times \bar{x} + 25\pi \times 20 \quad \longleftarrow \boxed{\text{Divide by } \pi.}$$

$$\Rightarrow \qquad \bar{x} = \frac{225 \times 15 - 25 \times 20}{200}$$

$$= 14.4$$

The centre of mass is 14.4 cm from O, that is, 0.6 cm to the left of the centre of the disc.

Centre of mass of a triangle

The triangle in Figure 3.16 is divided up into thin strips parallel to the side AB.

> The median of a triangle joins the vertex to the midpoint of the opposite side.

The centre of mass of each strip lies in the middle of the strip, at the points $C_1, C_2, C_3, \ldots$.

When these points are joined they form the median of the triangle, drawn from C.

Similarly, the centre of mass also lies on the medians from A and from B. Therefore, the centre of mass lies at the intersection of the three medians; this is the **centroid** of the triangle. This point is $\frac{2}{3}$ of the distance along the median from the vertex.

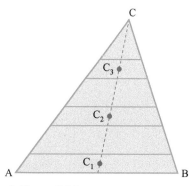

▲ **Figure 3.16**

Answers to exercises are available at www.hoddereducation.com/cambridgeextras

| Example 3.8 | Find the coordinates of the centre of mass of a uniform triangular lamina with vertices at A(4,4), B(1,1) and C(5,1). |

If the lamina is suspended from A, find the angle that the line BC makes with the horizontal.

Solution

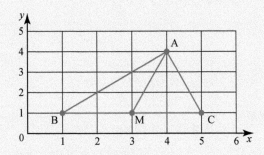

▲ Figure 3.17

The midpoint of BC is at M with coordinates (3,1). The centre of mass, G, is $\frac{2}{3}$ of the way down the median AM.

G has x coordinate: $4 - \frac{2}{3} \times (4 - 3) = 3\frac{1}{3}$

and y coordinate: $4 - \frac{2}{3} \times (4 - 1) = 2$

The centre of mass G of the lamina is at $\left(3\frac{1}{3}, 2\right)$.

The lamina is suspended from A. G will be directly below A. The line AG and also the median AM will be along the vertical.

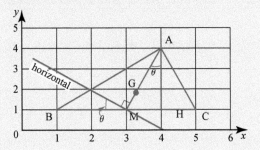

▲ Figure 3.18

The horizontal through M makes an angle θ with BC. The same angle θ is highlighted in the triangle AMH, from which you can derive

$$\tan\theta = \frac{MH}{AH} = \frac{4-3}{4-1} = \frac{1}{3} \Rightarrow \theta = 18.4°$$

The line BC makes an angle 18.4° with the horizontal when the lamina is suspended from A.

Centres of mass of other shapes

The table below gives the position of the centre of mass of some uniform objects that you may encounter, or wish to include within models of composite bodies. The centre of mass for each of these shapes lies on its line of symmetry. These standard results may be obtained using calculus methods.

Body	Position of centre of mass	Diagram
Solid cone or pyramid	$\frac{3}{4}h$ down from vertex	
Hollow cone or pyramid	$\frac{1}{3}h$ from centre of base	
Solid hemisphere	$\frac{3}{8}r$ up from centre of base	
Hemispherical shell	$\frac{1}{2}r$ up from centre of base	
Circular sector of radius r and angle 2α radians	$\dfrac{2r\sin\alpha}{3\alpha}$ from centre	
Circular arc of radius r and angle 2α radians	$\dfrac{r\sin\alpha}{\alpha}$ from centre	

Example 3.9

Find the centre of mass of the uniform lamina illustrated in Figure 3.19.

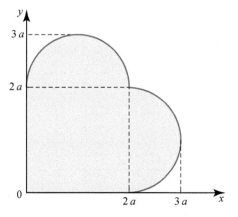

▲ **Figure 3.19**

Answers to exercises are available at www.hoddereducation.com/cambridgeextras

Solution

The shape is symmetric about the line $y = x$. It thus follows that the centre of mass lies on that line. In other words you need only find one of the coordinates. The shape is made up of a square of side $2a$ with two semicircles of radius a along two of the edges.

A semicircle of radius r subtends an angle π radians and so has a centre of mass located at $\dfrac{2r \sin \dfrac{\pi}{2}}{\dfrac{3\pi}{2}} = \dfrac{4r}{3\pi}$ from the centre of the semicircle.

The square has centre of mass at (a, a). The semicircles have centre of mass at $\left(a, 2a + \dfrac{4a}{3\pi}\right)$ and $\left(2a + \dfrac{4a}{3\pi}, a\right)$.

The mass of each shape is proportional to its area: $4a^2$ for the square and $\dfrac{1}{2}\pi a^2$ for the semicircles. This gives rise to the moment equation:

$$\left(4a^2 + \frac{1}{2}\pi a^2 + \frac{1}{2}\pi a^2\right)\bar{x} = 4a^2 \times a + \frac{1}{2}\pi a^2 \times a + \frac{1}{2}\pi a^2 \times \left(2a + \frac{4a}{3\pi}\right)$$

$$\bar{x}\left(4a^2 + \pi a^2\right) = \frac{14}{3}a^3 + \frac{3\pi}{2}a^3$$

$$\Rightarrow \bar{x} = \frac{a}{6}\left(\frac{28 + 9\pi}{4 + \pi}\right) = 1.31\ldots a.$$

The centre of mass is at $(1.31a, 1.31a)$.

Exercise 3B

1 Find the centre of mass of each set of point masses.

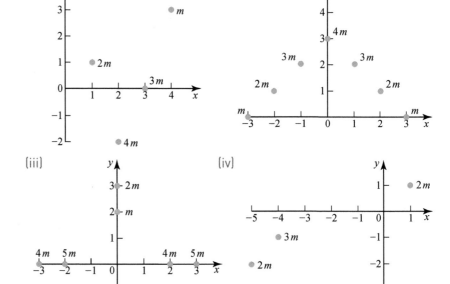

2 Masses of 1, 2, 3 and 4 grams are placed at the corners A, B, C and D of a square piece of uniform cardboard of side 10 cm and mass 5 g. Find the position of the centre of mass relative to axes through AB and AD.

3 As part of an illuminated display, letters are produced by mounting bulbs in holders 30 cm apart on light wire frames. The combined mass of a bulb and its holder is 200 grams. Find the position of the centre of mass for each of the letters shown below, in terms of its horizontal and vertical displacement from the bottom left-hand corner of the letter.

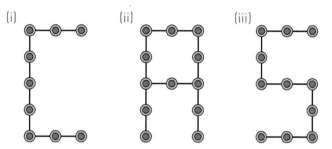

4 Four people of masses 60 kg, 65 kg, 62 kg and 75 kg sit on the four seats of the fairground ride shown below. The seats and the connecting arms are light. Find the radius of the circle described by the centre of mass when the ride rotates about O.

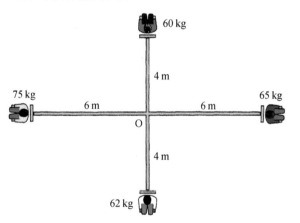

5 The following shapes are made out of uniform card.

For each shape, find the coordinates of the centre of mass relative to O.

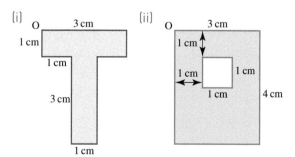

Answers to exercises are available at www.hoddereducation.com/cambridgeextras

6 Find the coordinates of the centre of mass of each of these uniform laminae.

(i)

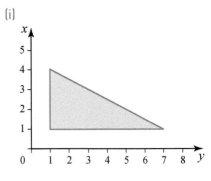

(ii)

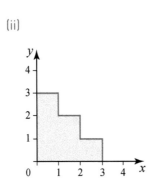

(iii)

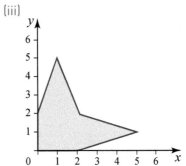

7 A pendant is made from a uniform circular disc of mass $4m$ and radius 2 cm with a decorative edging of mass m, as shown in the diagram. The centre of mass of the decoration is 1 cm below the centre, O, of the disc. The pendant is symmetrical about the diameter AB.

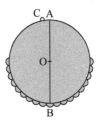

(i) Find the position of the centre of mass of the pendant.

The pendant should be hung from A but the light ring for hanging it is attached at C, where angle AOC is 10°.

(ii) Find the angle between AB and the vertical when the pendant is hung from C.

8 ABCD is a rectangular plate, with AB = 5 cm and AD = 8 cm. E is the midpoint of BC. The triangular portion ABE is removed and the remainder is suspended from A. Find the angle that AD makes with the vertical.

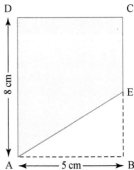

9 A uniform rectangular lamina, ABCD, where AB is of length $2a$ and BC of length $5a$, has a mass $20m$. Further point masses m, $2m$, $3m$ and $4m$ are fixed to the points A, B, C and D, respectively.

    (i)   Find the centre of mass of the system, in terms of a, relative to x- and y-axes along AB and AD, respectively.

    (ii)   If the lamina is suspended from the point A, find the angle that the diagonal AC makes with the vertical.

    (iii) What magnitude must the mass at point D have if AC is to hang vertically?

PS

10 A vase is made from a uniform solid cylinder of height 25 cm and diameter 10 cm by removing a smaller cylinder of height 22 cm and diameter 9 cm from it so that there is an axis of symmetry vertically through the centre of the base.

Find the centre of mass of the vase.

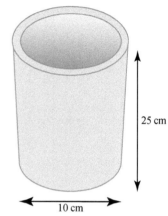

25 cm

10 cm

11 A uniform lamina ABC is in the form of a major segment of a circle with centre O and radius 0.35 m. The straight edge of the lamina is AB, and angle AOB $= \frac{2}{3}\pi$ radians (see diagram).

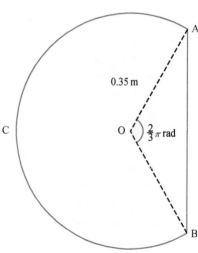

0.35 m

O

$\frac{2}{3}\pi$ rad

A

B

C

    (i)   Show that the centre of mass of the lamina is 0.0600 m from O, correct to 3 significant figures.

The weight of the lamina is 14 N. It is placed on a rough horizontal surface with A vertically above B and the lowest point of the arc BC in contact with the surface. The lamina is held in equilibrium in a vertical plane by a force of magnitude F N acting at A.

    (ii)   Find F in each of the following cases:

        (a)   the force of magnitude F N acts along AB;

        (b)   the force of magnitude F N acts along the tangent to the circular arc at A.

Cambridge International AS & A Level Mathematics
9709 Paper 53 Q7 November 2014

Answers to exercises are available at www.hoddereducation.com/cambridgeextras

12 The diagram shows a circular object formed from a uniform semi-circular lamina of weight 11 N and a uniform semi-circular arc of weight 9 N. The lamina and arc both have centre O and radius 0.7 m and are joined at the ends of their common diameter AB.

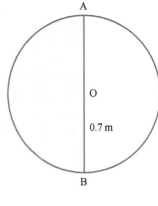

(i) Show that the distance of the centre of mass of the object from O is 0.0371 m correct to 3 significant figures.

The object hangs in equilibrium, freely suspended at A.

(ii) Find the angle between AB and the vertical and state whether the lowest point of the object is on the lamina or on the arc.

Cambridge International AS & A Level Mathematics
9709 Paper 51 Q2 June 2012

CP

13 A toy consists of a solid hemisphere of radius R joined to a solid right cone of radius R and height H. The solids are joined at their flat surfaces. The hemispherical base is made of a material that is twice as dense as the conical top.

Show that the centre of mass of the toy lies at a distance

$$\frac{H^2 - 6R^2}{4(H + 4R)}$$

from the common face of the two solids.

PS

14 Find the centre of mass of the following uniform lamina.

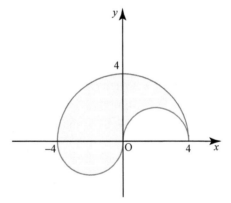

CP

15 A drink can is symmetrical, with height h cm, and when empty its mass is m grams. The drink that fills it has mass M grams and can be taken to fill the can completely.

(i) Find the height of the centre of mass when the can is standing on a level table and

(a) is half full

(b) a proportion, α, of the drink remains.

When the can is full the centre of mass is clearly halfway up it, at a height of $\frac{1}{2}h$. The same is true when it is completely empty. In between these two extremes, the centre of mass is below the middle.

(ii) Show that, when the centre of mass is at its minimum height, α satisfies the equation $M\alpha^2 + 2\alpha m - m = 0$ and that the centre of mass lies on the surface of the drink.

3.3 Sliding and toppling

The photograph shows a double-decker bus on a test ramp. The angle of the ramp to the horizontal is slowly increased.

▲ **Figure 3.20**

[Photo courtesy of Millbrook Proving Ground Ltd]

> ❯ What happens to the bus in the photograph? Would a loaded bus behave differently from the empty bus shown here?

EXPERIMENT

The diagrams in Figure 3.21 show a force being applied in different positions to a cereal packet.

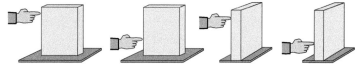

▲ **Figure 3.21**

In which case do you think the packet is most likely to fall over? In which case is it most likely to slide? Investigate your answers practically, using boxes of different shapes.

Answers to exercises are available at www.hoddereducation.com/cambridgeextras

> Figure 3.22 shows a cereal packet placed on a slope. Is the box more likely to topple or slide as the angle of the slope to the horizontal increases?

▲ **Figure 3.22**

> To what extent is this situation comparable to that of the bus on the test ramp in Figure 3.20?

Two critical cases

When an object stands on a surface, the only forces acting are its weight W and the *resultant* of all the contact forces between the surfaces, which must act through a point on both surfaces. This resultant contact force is often resolved into two components: the friction, F, parallel to any possible sliding, and the normal reaction, R, perpendicular to F, as in Figures 3.23–3.25.

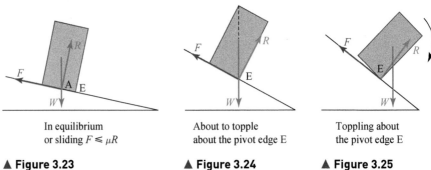

In equilibrium or sliding $F \leqslant \mu R$	About to topple about the pivot edge E	Toppling about the pivot edge E
▲ **Figure 3.23**	▲ **Figure 3.24**	▲ **Figure 3.25**

Equilibrium can be broken in two ways:

1 *The object is on the point of sliding*; then $F = \mu R$ according to our model.

2 *The object is on the point of toppling.* The pivot is at the lowest point of contact, which is the point E in Figure 3.24. In this critical case:

   (i) the centre of mass is directly above E so the weight acts vertically downwards through E;

   (ii) the resultant reaction of the plane on the object acts through E, vertically upwards. This is the resultant of F and R.

> Why does the object topple in Figure 3.25?

When three non-parallel forces are in equilibrium, their lines of action must be concurrent (they must all pass through one point). Otherwise there is a resultant moment about the point where two of them meet, as in Figure 3.25.

Example 3.10

An increasing force PN is applied to a block, as shown in Figure 3.26, until the block moves. The coefficient of friction between the block and the plane is 0.4. Does it slide or topple?

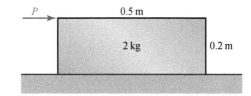

▲ **Figure 3.26**

Solution

The forces acting are shown in Figure 3.27. The normal reaction may be thought of as a single force acting somewhere within the area of contact. When toppling occurs (or is about to occur), the line of action is through the edge about which it topples.

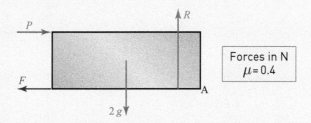

Forces in N
$\mu = 0.4$

▲ **Figure 3.27**

Until the block moves, it is in equilibrium.

Horizontally:	$P = F$	①
Vertically:	$R = 2g$	②
If *sliding* is about to occur	$F = \mu R$	
From ①	$P = \mu R = 0.4 \times 2g$	
	$= 8$	

If the block is about to *topple*, then A is the pivot point and the reaction of the plane on the block acts at A. Taking moments about A gives

$$(\curvearrowleft) \quad 2g \times 0.25 - P \times 0.2 = 0$$
$$P = 25$$

R acts through A.

So, in order for the block to slide, P needs to exceed 8 N, but for the block to topple, it needs to exceed 25 N: the block will slide before it topples.

Answers to exercises are available at www.hoddereducation.com/cambridgeextras

Example 3.11

A rectangular block of mass 3 kg is placed on a slope as shown. The angle α is gradually increased. What happens to the block, given that the coefficient of friction between the block and the slope is 0.6?

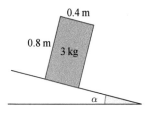

▲ **Figure 3.28**

Solution

Check for possible sliding

Figure 3.29 shows the forces acting when the block is in equilibrium.

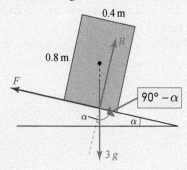

▲ **Figure 3.29**

Resolve parallel to the slope: $\qquad F = 3g \sin \alpha$
Perpendicular to the slope: $\qquad R = 3g \cos \alpha$

When the block is on the point of sliding $F = \mu R$, so

$$3g \sin \alpha = \mu \times 3g \cos \alpha$$
$$\Rightarrow \qquad \tan \alpha = \mu = 0.6$$
$$\Rightarrow \qquad \alpha = 31.0°$$

The block is on the point of sliding when $\alpha = 31.0°$.

Check for possible toppling

When the block is on the point of toppling about the edge E, the centre of mass is vertically above E, as shown in Figure 3.30.

Then the angle α is given by:
$$\tan \alpha = \frac{0.4}{0.8}$$
$$\alpha = \arctan(0.5) = 26.6°$$

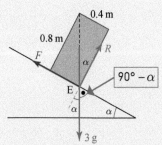

▲ **Figure 3.30**

The block topples when $\alpha = 26.6°$.

The angle for sliding (31.0°) is greater than the angle for toppling (26.6°), so the block topples without sliding when $\alpha = 26.6°$.

> ? Is it possible for sliding and toppling to occur for the same angle?

Exercise 3C

1 A force of magnitude P N acts as shown on a block resting on a horizontal plane. The coefficient of friction between the block and the plane is 0.7.

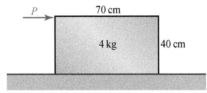

The magnitude of the force P is gradually increased from zero.

(i) Find the magnitude of P if the block is on the point of sliding, assuming it does not topple.

(ii) Find the magnitude of P if the block is on the point of toppling, assuming it does not slide.

(iii) Does the block slide or topple?

2 A solid uniform cuboid is placed on a horizontal surface. A force P is applied as shown in the diagram.

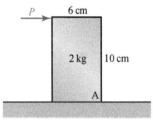

(i) The block is on the point of sliding. Express P in terms of μ, the coefficient of friction between the block and the plane.

(ii) Find the magnitude of P if the cuboid is on the point of toppling.

(iii) For what values of μ will the block slide before it topples?

(iv) For what values of μ will the block topple before it slides?

3 A uniform rectangular block of height 30 cm and width 10 cm is placed on a rough plane inclined at an angle α to the horizontal. The box lies on the plane with its length horizontal. The coefficient of friction between the block and the plane is 0.25.

(i) Assuming that it does not topple, for what value of α does the block begin to slide?

(ii) Assuming that it does not slide, for what value of α does the block begin to topple?

(iii) The angle α is increased slowly from 0. Which happens first, sliding or toppling?

4   A horizontal force of increasing magnitude is applied to the middle of the face of a 50 cm uniform cube, at right angles to the face. The coefficient of friction between the cube and the surface is 0.4 and the cube is on a level surface. What happens to the cube?

5   A solid uniform cube of side 4 cm and weight 60 N is situated on a rough horizontal plane. The coefficient of friction between the cube and the plane is 0.4. A force P N acts in the middle of one of the edges of the top of the cube, as shown in the diagram.

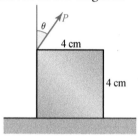

In the cases when the value of θ is (a) 60° and (b) 80°, find

(i)   the force P needed to make the cube slide, assuming it does not topple

(ii)   the force P needed to make the cube topple, assuming it does not slide

(iii)   whether it first slides or topples as the force P is increased.

(iv)   For what value of θ do toppling and sliding occur for the same value of P, and what is that value of P?

6   A solid uniform cuboid, 10 cm × 20 cm × 50 cm, is to stand on an inclined plane that makes an angle α with the horizontal. One edge of the cuboid is to be parallel to the line of the slope. The coefficient of friction between the cuboid and the plane is μ.

(i)   Which face of the cuboid should be placed on the slope to make it

(a)   least likely   (b)   most likely to topple?

(ii)   How does the cuboid's orientation influence the likelihood of it sliding?

(iii)   Find the range of values of μ in the situations where

(a)   it will slide first whatever its orientation

(b)   it will topple first whatever its orientation.

7   A man is trying to move a uniform scaffold plank ABCD of length 2.5 m and weight 120 N that is resting on horizontal ground. You may assume that he exerts a slowly increasing force of magnitude P N at a constant angle θ to the vertical and at right angles to the edge CD, as shown in the diagram.

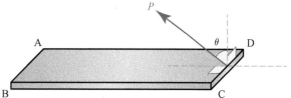

As P increases, the plank will either slide or start to turn about the end AB, depending on the values of θ and the coefficient of friction, μ, between the plank and the ground.

Assume that the plank slides before it turns and is on the point of sliding.

(i)   Show that the normal reaction of the ground on the plank is $(120 - P\cos\theta)$ N.

(ii)   Obtain two expressions involving the frictional force acting on the plank and deduce that

$$P = \frac{120\mu}{\sin\theta + \mu\cos\theta}$$

Assume now that the plank starts to turn about the edge AB before it slides and is on the point of turning.

(iii)   Where is the line of action of the normal reaction of the ground on the plank?

(iv)   Show that $P = \dfrac{60}{\cos\theta}$.

(v)   Given that the plank slides before it turns about AB as the force P is gradually increased, find the relationship between μ and θ.

Answers to exercises are available at www.hoddereducation.com/cambridgeextras

8   A uniform solid is made from a cylinder
    and a cone, both with radius 0.5 m and
    height 0.4 m.

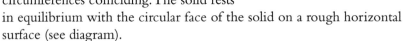

    The circular base of the cone is attached
    to a circular face of the cylinder, with their
    circumferences coinciding. The solid rests
    in equilibrium with the circular face of the solid on a rough horizontal
    surface (see diagram).

    (i)   Show that the centre of mass of the solid is 0.275 m above
          the surface.

    The weight of the solid is 60 N. A horizontal force of increasing
    magnitude P N is applied to the vertex of the cone, which causes the
    solid eventually to topple without sliding.

    (ii) Calculate the value of P for which the solid is on the point
          of toppling.

    (iii) Find the least possible value for the coefficient of friction between
          the solid and the surface.

    The force of magnitude P is removed, and the solid is held with the
    curved surface of the cylinder in contact with the horizontal surface.
    The horizontal surface is then tilted so that it makes an angle of 30°
    with the horizontal. The solid is released with its axis of symmetry
    parallel to a line of greatest slope and the conical portion pointing down
    the slope.

    (iv) Show that the solid does not slide, but does topple.

Cambridge International AS & A Level Mathematics
9709 Paper 53 Q7 November 2013

9   A filing cabinet has the dimensions shown in the diagram. The body of
    the cabinet has mass 20 kg and its construction is such that its centre of
    mass is at a height of 60 cm, and is 25 cm from the back of the cabinet.
    The mass of a drawer and its contents may be taken to be 10 kg and its
    centre of mass to be 10 cm above its base and 30 cm from its front face.

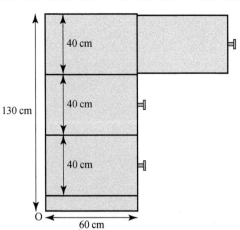

    (i)   Find the position of the centre of mass when all the drawers are closed.

(ii) Find the position of the centre of mass when the top two drawers are fully open.

(iii) Show that when all three drawers are fully opened the filing cabinet will tip over.

(iv) Two drawers are fully open. How far can the third one be opened without the cabinet tipping over?

CP **10** Uniform wooden bricks have length 20 cm and height 5 cm. They are glued together as shown in the diagram, with each brick 5 cm to the right of the one below it. The origin is taken to be at O.

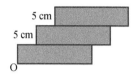

(i) Find the coordinates of the centre of mass for

(a) 1 (b) 2 (c) 3 (d) 4 (e) 5 bricks.

(ii) How many bricks can be assembled in this way without them tipping over?

(iii) If the displacement were changed from 5 cm to 2 cm, find the coordinates of the centre of mass for *n* bricks. How many bricks can now be assembled without them tipping over?

(iv) If the displacement is $\frac{1}{2}$ cm, what is the maximum height possible for the centre of mass of such an assembly of bricks without them tipping over?

PS **11** A uniform solid cylinder has radius 0.7 m and height *h* m. A uniform solid cone has base radius 0.7 m and height 2.4 m. The cylinder and the cone both rest in equilibrium, each with a circular face in contact with a horizontal plane. The plane is now tilted so that its inclination to the horizontal, θ, is increased gradually until the cone is about to topple.

(i) Find the value of θ at which the cone is about to topple.

(ii) Given that the cylinder does not topple, find the greatest possible value of *h*.

The plane is returned to a horizontal position and the cone is fixed to one end of the cylinder so that the plane faces coincide. It is given that the weight of the cylinder is three times the weight of the cone. The curved surface of the cone is placed on the horizontal plane (see diagram).

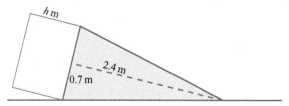

(iii) Given that the solid immediately topples, find the least possible value of *h*.

12 A cube of side 4 cm and mass 100 g is acted on by a force, as shown in the diagram.

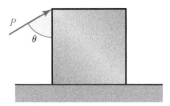

The coefficient of friction between the cube and the plane is 0.3. What happens to the cube if

(i) $\theta = 45°$ and $P = 0.3\,\text{N}$?

(ii) $\theta = 15°$ and $P = 0.45\,\text{N}$?

13 A cube of side l and weight W is resting on a rough horizontal plane. A force of magnitude T is applied to a point P on the top edge of the cube. This force makes an angle θ with the horizontal such that $\tan\theta = \dfrac{3}{4}$ and the force is in the vertical plane containing P and the centre of the cube. The cube does not slip and is on the point of tipping.

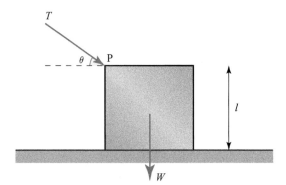

Show that $T = 2.5W$.

KEY POINTS

1   The centre of mass of a body has the property that the moment, about any point, of the whole mass of the body taken at the centre of mass is equal to the sum of the moments of the various particles comprising the body.

$$M\overline{\mathbf{r}} = \sum m_i\mathbf{r}_i \text{ where } M = \sum m_i$$

2   In one dimension

$$M\overline{x} = \sum m_i x_i$$

3   In two dimensions

$$M\begin{pmatrix} \overline{x} \\ \overline{y} \end{pmatrix} = \sum m_i \begin{pmatrix} x_i \\ y_i \end{pmatrix}$$

4   In three dimensions

$$M\begin{pmatrix} \overline{x} \\ \overline{y} \\ \overline{z} \end{pmatrix} = \sum m_i \begin{pmatrix} x_i \\ y_i \\ z_i \end{pmatrix}$$

LEARNING OUTCOMES

Now that you have finished this chapter, you should be able to
- find the centre of mass of a system of particles, each with a given position and mass
- find the centre of mass of a simple shape
- use symmetry when finding a centre of mass
- find or recall the positions of the centres of mass of simple shapes
- find the centre of mass of a composite body
- use centre of mass in problems involving the equilibrium of a rigid body.

4 Circular motion

> ❯ These photographs show some objects that move in circular paths. What other examples can you think of?
> ❯ What makes objects move in circles?
> ❯ Why does the Moon circle the Earth?
> ❯ What happens to the 'hammer' when the athlete lets it go?
> ❯ Do the pilots of the planes need to be strapped into their seats at the top of a loop in order not to fall out?

The answers to these questions lie in the nature of circular motion. Even if an object is moving at constant speed in a circle, its velocity keeps changing because its direction of motion keeps changing. Consequently the object is accelerating and so, according to Newton's first law, there must be a force acting on it. The force required to keep an object moving in a circle can be provided in many ways.

Without the Earth's gravitational force, the Moon would move off at constant speed in a straight line into space. The wire attached to the athlete's hammer provides a tension force that keeps the ball moving in a circle. When the athlete lets go, the ball flies off at a tangent because the tension has disappeared.

Although it would be sensible for the pilots to be strapped in, no upward force is necessary to stop them falling out of the plane because their weight contributes to the force required for motion in a circle.

In this chapter, these effects are explained.

4.1 Notation

To describe circular motion (or indeed any other topic) mathematically you need suitable notation. It will be helpful in this chapter to use the notation (attributed to Newton) for differentiation with respect to time in which, for example, $\frac{ds}{dt}$ is written as $\dot{s}$, and $\frac{d^2\theta}{dt^2}$ as $\ddot{\theta}$.

Figure 4.1 shows a particle P moving round the circumference of a circle of radius r, centre O. At time t, the position vector $\overrightarrow{OP}$ of the particle makes an angle θ (in radians) with the fixed direction $\overrightarrow{OA}$. The arc length AP is denoted by s.

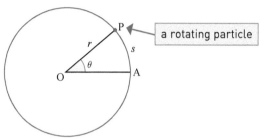

a rotating particle

▲ **Figure 4.1**

4.2 Angular speed

Using this notation

$$s = r\theta$$

Differentiating this with respect to time (using the product rule) gives:

$$\frac{ds}{dt} = r\frac{d\theta}{dt} + \theta\frac{dr}{dt}.$$

Since r is constant for a circle, $\frac{dr}{dt} = 0$, so the rate at which the arc length increases is

$$\frac{ds}{dt} = r\frac{d\theta}{dt} \quad \text{or} \quad \dot{s} = r\dot{\theta}. \qquad \text{①}$$

In this equation $\dot{s}$ is the speed at which P is moving round the circle (often denoted by v), and $\dot{\theta}$ is the rate at which the angle θ is increasing, which is the rate at which the position vector $\overrightarrow{OP}$ is rotating.

The quantity $\frac{d\theta}{dt}$, or $\dot{\theta}$, can be called the **angular velocity** or the **angular speed** of P. In more advanced work, angular velocity is treated as a vector and its direction is taken to be that of the axis of rotation. In this book, $\frac{d\theta}{dt}$ is often referred to as angular speed, but is given a sign: positive when θ is increasing (usually anticlockwise) and negative when θ is decreasing (usually clockwise).

Answers to exercises are available at www.hoddereducation.com/cambridgeextras

Angular speed is often denoted by ω, the Greek letter omega. So the equation $\dot{s} = r\dot{\theta}$ may be written as

$$v = r\omega.$$

Notice that for this equation to hold, θ must be measured in radians, so the angular speed is measured in **radians per second** or $\text{rad}\,\text{s}^{-1}$.

> It is common practice to give angular speed as a multiple of π.

Figure 4.2 shows a disc rotating about its centre, O, with angular speed ω. The line OP represents any radius.

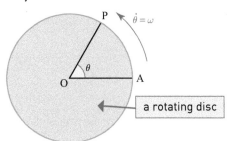

P $\quad \dot{\theta} = \omega$

θ

O $\qquad$ A

a rotating disc

▲ **Figure 4.2**

Every point on the disc describes a circular path, and all points have the same angular speed. However the *actual* speed of any point depends on its distance from the centre: increasing r in the equation $v = r\omega$ increases v. You will appreciate this if you have ever been at the end of a rotating line of people in a dance or watched a body of marching soldiers wheeling round a corner.

Angular speeds are sometimes measured in revolutions per second or revolutions per minute (rpm) where one revolution is equal to 2π radians. For example, turntables for vinyl records may rotate at 78, 45 or $33\frac{1}{3}$ rpm, while a computer hard disc might spin at 7200 rpm or more; at cruising speeds, crankshafts in car engines typically rotate at 3000 to 4000 rpm.

Example 4.1

A police car drives at $64\,\text{km}\,\text{h}^{-1}$ around a circular bend of radius $16\,\text{m}$. A second car moves so that it has the same angular speed as the police car but in a circle of radius $12\,\text{m}$. Is the second car breaking the $50\,\text{km}\,\text{h}^{-1}$ speed limit?

Solution

Converting kilometres per hour to metres per second gives

$$64\,\text{km}\,\text{h}^{-1} = \frac{64 \times 1000}{3600}\,\text{m}\,\text{s}^{-1}$$

$$= \frac{160}{9}\,\text{m}\,\text{s}^{-1}$$

Using $v = r\omega$, $\qquad \omega = \frac{160}{9 \times 16}\,\text{rad}\,\text{s}^{-1}$

$$= \frac{10}{9}\,\text{rad}\,\text{s}^{-1}$$

The speed of the second car is
$$v = 12\omega$$
$$= \frac{10}{9} \times 12\,\mathrm{m\,s^{-1}}$$
$$= \frac{120 \times 3600}{9 \times 1000}\,\mathrm{km\,h^{-1}}$$
$$= 48\,\mathrm{km\,h^{-1}}$$

The second car is just below the speed limit.

 Notes

1 Notice that working in fractions gives an exact answer.

2 A quicker way to solve this problem would be to notice that, because the cars have the same angular speed, the actual speeds of the cars are proportional to the radii of the circles in which they are moving. Using this method it is possible to stay in $\mathrm{km\,h^{-1}}$. The ratio of the two radii is $\frac{12}{16}$ so the speed of the second car is $\frac{12}{16} \times 64\,\mathrm{km\,h^{-1}} = 48\,\mathrm{km\,h^{-1}}$.

Exercise 4A

1 Find the angular speed, in radians per second correct to one decimal place, of records rotating at

(i)   78 rpm

(ii)   45 rpm

(iii) $33\frac{1}{3}$ rpm.

2 A flywheel is rotating at $300\,\mathrm{rad\,s^{-1}}$. Express this angular speed in rpm, correct to the nearest whole number.

3 The Star of Nanchang is a giant observation ferris wheel with a diameter of 153 m. Each one of the 60 observation cabins completes one revolution in 30 minutes.

(i)   Calculate the angular speed in

(a)   rpm

(b)   radians per second.

(ii)   Calculate the speed of the point on the circumference where passengers board the moving wheel.

4 A lawnmower engine is started by pulling a rope that has been wound round a cylinder of radius 4 cm. Find the angular speed of the cylinder at a moment when the rope is being pulled with a speed of $1.3\,\mathrm{m\,s^{-1}}$. Give your answer in radians per second, correct to one decimal place.

5 The wheels of a car have radius 20 cm. What is the angular speed, in radians per second correct to one decimal place, of a wheel when the car is travelling at

(i) $10\,\mathrm{m\,s^{-1}}$ (ii) $30\,\mathrm{m\,s^{-1}}$?

Answers to exercises are available at www.hoddereducation.com/cambridgeextras

6 The angular speed of an audio CD changes continuously, so that a laser can read the data at a constant speed of $12\,\mathrm{m\,s^{-1}}$. Find the angular speed (in rpm) when the distance of the laser from the centre is

    (i) $30\,\mathrm{mm}$ (ii) $55\,\mathrm{mm}$.

M

7 What is the average angular speed of the Earth in radians per second as it

    (i)   orbits the Sun?   (ii)   rotates about its own axis?

The radius of the Earth is $6400\,\mathrm{km}$.

    (iii)   At what speed is someone on the equator travelling relative to the centre of the Earth?

    (iv)   Hamid lives in Pabna in Bangladesh; the latitude there is $24°\,\mathrm{N}$. At what speed does he travel relative to the centre of the Earth? Give your answer in $\mathrm{km\,h^{-1}}$ to the nearest $10\,\mathrm{km\,h^{-1}}$.

8 A tractor has front wheels of diameter $70\,\mathrm{cm}$ and back wheels of diameter $1.6\,\mathrm{m}$. What is the ratio of their angular speeds when the tractor is being driven along a straight road?

9 (i)   Find the kinetic energy of a $50\,\mathrm{kg}$ person riding a big wheel with radius $5\,\mathrm{m}$ when the ride is rotating at $3\,\mathrm{rpm}$. You should assume that the person can be modelled as a particle.

    (ii)   Explain why this modelling assumption is necessary.

10 The minute hand of a clock is $1.2\,\mathrm{m}$ long and the hour hand is $0.8\,\mathrm{m}$ long.

    (i)   Find the speeds of the tips of the hands.

    (ii)   Find the ratio of the speeds of the tips of the hands and explain why this is not the same as the ratio of the angular speeds of the hands.

11 The diagram represents a 'Chairoplane' ride at a fair. It completes one revolution every 2.5 seconds.

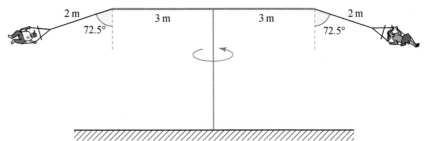

    (i)   Find the radius of the circular path that a rider follows.

    (ii)   Find the speed of a rider.

CP

12 The position vector of a rider on a helter-skelter is given by

$$\mathbf{r} = 2\sin t\,\mathbf{i} + 2\cos t\,\mathbf{j} + \left(8 - \tfrac{1}{2}t\right)\mathbf{k}$$

where the units are in metres and seconds. The unit vector $\mathbf{k}$ acts vertically upwards.

    (i)   Find an expression for the velocity of the rider at time t.

    (ii)   Find the speed of the rider at time t.

    (iii)   Find the magnitude and direction of the rider's acceleration when $t = \dfrac{\pi}{4}$.

4.3 Velocity and acceleration

Velocity and acceleration are both vector quantities. They can be expressed either in magnitude–direction form or in components. When describing circular motion or other orbits it is most convenient to take components in directions along the radius (**radial direction**) and at right angles to it (**transverse direction**).

For a particle moving round a circle of radius r, the velocity has:

radial component: 0

transverse component: $r\dot{\theta}$ or $r\omega$

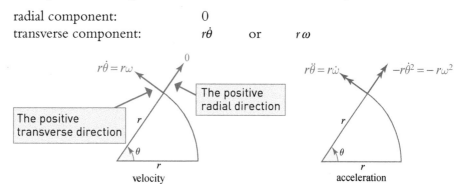

▲ **Figure 4.3**   ▲ **Figure 4.4**

The acceleration of a particle moving round a circle of radius r has:

radial component: $-r\dot{\theta}^2$ or $-r\omega^2$

transverse component: $r\ddot{\theta}$ or $r\dot{\omega}$.

The transverse component is just what you would expect: the radius multiplied by the angular acceleration, $\ddot{\theta}$. If the particle has constant angular speed, its angular acceleration is zero and so the transverse component of its acceleration is also zero.

In contrast, the radial component of the acceleration, $-r\omega^2$, is almost certainly not a result you would have expected intuitively. It tells you that a particle travelling in a circle is always accelerating towards the centre of the circle, but without ever getting any closer to the centre. If this seems a strange idea, you may find it helpful to remember that circular motion is not a natural state; left to itself a particle will travel in a straight line. To keep a particle in the unnatural state of circular motion it must be given an acceleration at right angles to its motion, which is towards the centre of the circle.

Circular motion with constant speed

In this section, the circular motion is assumed to be uniform and thus have no transverse component of acceleration. Later in the chapter, you will study situations where the angular speed varies.

Problems involving circular motion often refer to the actual speed of the object, rather than its angular speed. It is easy to convert the one into the other using the relationship $v = r\omega$.

The relationship $v = r\omega$ can also be used to express the magnitude of the acceleration in terms of v and r:

$$\omega = \frac{v}{r}$$

$$a = r\omega^2 = r\left(\frac{v}{r}\right)^2$$

$$\Rightarrow \quad a = \frac{v^2}{r} \text{ towards the centre.}$$

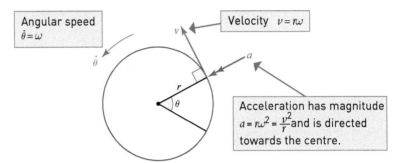

▲ **Figure 4.5**

Example 4.2

A turntable is rotating at 45 rpm. A fly is standing on the turntable at a distance of 8 cm from the centre. Find

(i)   the angular speed of the fly in radians per second

(ii) the speed of the fly in metres per second

(iii) the acceleration of the fly.

Solution

(i) $45 \text{ rpm} = 45 \times 2\pi \text{ rad min}^{-1}$ ← One revolution is 2π radians.

$$= \frac{45 \times 2\pi}{60} \text{ rad s}^{-1}$$

$$= \frac{3\pi}{2} \text{ rad s}^{-1}.$$

(ii) If the speed of the fly is $v \text{ m s}^{-1}$, v can be found using

$$v = r\omega$$

$$= 0.08 \times \frac{3\pi}{2}$$

$$= 0.377$$

So the speed of the fly is 0.377 m s^{-1}.

(iii) The acceleration of the fly is given by

$$r\omega^2 = 0.08 \times \left(\frac{3\pi}{2}\right)^2$$

$$= 1.78$$

The acceleration of the fly is 1.78 m s^{-2} directed towards the centre of the turntable.

4.4 The forces required for circular motion

Newton's first law of motion states that a body will continue in a state of rest or uniform motion in a straight line unless acted upon by an external force. Any object moving in a circle, such as the police car and the fly in Examples 4.1 and 4.2, must therefore be acted upon by a resultant force in order to produce the required acceleration towards the centre.

A force towards the centre is called a **centripetal** (centre-seeking) force. A resultant centripetal force is necessary for a particle to move in a circular path.

4.5 Circular motion in a horizontal circle

You are now in a position to use Newton's second law to determine theoretical answers to some of the questions that were posed at the beginning of this chapter. These will, as usual, be obtained using models of the true motion, which will be based on simplifying assumptions, for example zero air resistance. Large objects are assumed to be particles concentrated at their centres of mass.

Example 4.3

A coin is placed on a rotating turntable. Its centre is 5 cm from the centre of rotation. The coefficient of friction, μ, between the coin and the turntable is 0.5.

(i)   The speed of rotation of the turntable is gradually increased. At what angular speed will the coin begin to slide?

(ii) What happens next?

Solution

(i)   Because the speed of the turntable is increased only gradually, it can be assumed that the coin will not slip tangentially.

Figure 4.6 shows the forces acting on the coin, and its acceleration.

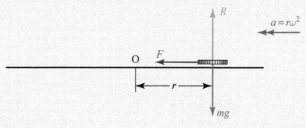

▲ **Figure 4.6**

The acceleration is towards the centre, O, of the circular path, so there must be a frictional force F in that direction.

There is no vertical component of acceleration, so the resultant force acting on the coin has no vertical component.

Therefore
$$R - mg = 0$$
$$R = mg \qquad \qquad \text{①}$$

Using Newton's second law for motion towards the centre of the circle:
$$F = ma = mr\omega^2 \qquad \qquad \text{②}$$

The coin will not slide, as long as $F \le \mu R$.

Substituting from ② and ① this gives

> Notice that the mass, m, has been eliminated at this stage, so that the answer does not depend upon it.

$$mr\omega^2 \le \mu mg$$

$$\Rightarrow \qquad r\omega^2 \le \mu g \longleftarrow$$

Taking g as $10\,\mathrm{m\,s^{-2}}$ and substituting $r = 0.05$ and $\mu = 0.5$

$$\omega^2 \le 100$$

$$\omega \le 10$$

The coin will move in a circle provided that the angular speed is less than $10\,\mathrm{rad\,s^{-1}}$, and this speed is independent of the mass of the coin.

(ii) When the angular speed increases beyond this, the coin slips to a new position, which is further from the centre of the circle. If the angular speed continues to increase, the coin will slip right off the turntable. When it reaches the edge it will fly off in the direction of the tangent.

The conical pendulum

A conical pendulum consists of a small bob tied to one end of a string. The other end of the string is fixed and the bob is made to rotate in a horizontal circle below the fixed point so that the string describes a cone, as shown in Figure 4.7.

▲ Figure 4.7

EXPERIMENT

1   Draw a diagram showing the magnitude and direction of the acceleration of a bob and the forces acting on it.

2   In the case that the radius of the circle remains constant, try to predict the effect on the angular speed when the length of the string is increased or when the mass of the bob is increased. What might happen when the angular speed increases?

3   Draw two circles of equal diameter on horizontal surfaces so that two people can make the bobs of conical pendulums rotate in circles of the same radius.

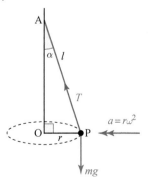

▲ **Figure 4.8**

(i)   Compare pendulums of different lengths with bobs of equal mass.

(ii) Compare pendulums of the same length but with bobs of different masses.

(iii) Does the angular speed depend on the length of the pendulum or the mass of the bob?

4   What happens when somebody makes the speed of the bob increase?

5   Can a bob be made to rotate with the string horizontal?

Theoretical model for the conical pendulum

A conical pendulum may be modelled as a particle of mass m attached to a light, inextensible string of length l. The mass is rotating in a horizontal circle with angular speed ω and the string makes an angle α with the downward vertical. The radius of the circle is r and the tension in the string is T, all in consistent units (S.I. units). The situation is shown in Figure 4.9.

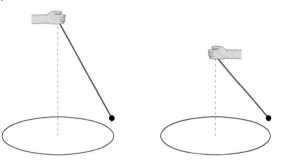

▲ **Figure 4.9**

Answers to exercises are available at www.hoddereducation.com/cambridgeextras

The magnitude of the acceleration is $r\omega^2$. The acceleration acts in a horizontal direction towards the centre of the circle. This means that there must be a resultant force acting towards the centre of the circle.

There are two forces acting on this particle, its weight mg and the tension T in the string.

As the acceleration of the particle has no vertical component, the resultant force has no vertical component, so

$$T\cos\alpha - mg = 0 \qquad \qquad \text{①}$$

Using Newton's second law towards the centre, O, of the circle

$$T\sin\alpha = ma = mr\omega^2 \qquad \qquad \text{②}$$

In triangle AOP

$$r = l\sin\alpha$$

Substituting for r in ② gives

$$T\sin\alpha = m(l\sin\alpha)\omega^2$$

$$\Rightarrow \quad T = ml\omega^2$$

Substituting this in ① gives

$$ml\omega^2\cos\alpha - mg = 0$$

$$\Rightarrow \qquad l\cos\alpha = \frac{g}{\omega^2} \qquad \qquad \text{③}$$

This equation provides sufficient information to give theoretical answers to the questions in the experiment.

» When r is kept constant and the length of the string is increased, the length AO $= l\cos\alpha$ increases. Equation ③ indicates that the value of $\frac{g}{\omega^2}$ increases and so the angular speed ω decreases. Conversely, the angular speed increases when the string is shortened.

» The mass of the particle does not appear in equation ③, so it has no effect on the angular speed, ω.

» When the length of the pendulum is unchanged, but the angular speed is increased, $\cos\alpha$ decreases, leading to an increase in the angle α and hence in r.

» If $\alpha \geqslant 90°$, $\cos\alpha \leqslant 0$, so $\frac{g}{\omega^2} \leqslant 0$, which is impossible. You can see from Figure 4.9 on page 95 that the tension in the string must have a vertical component to balance the weight of the particle.

Example 4.4

In Figure 4.10, the diagram on the right represents one of several arms of a fairground ride, shown on the left. The arms rotate about an axis and riders sit in chairs linked to the arms by chains.

4

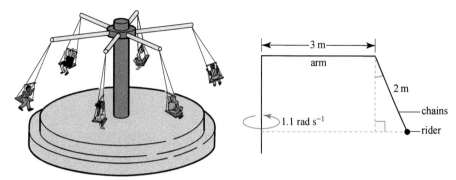

▲ **Figure 4.10**

The chains are 2 m long and the arms are 3 m long. Find the angle that the chains make with the vertical when the rider rotates at $1.1\,\text{rad}\,\text{s}^{-1}$.

Solution

Let $T\,\text{N}$ be the resultant tension in the chains holding a chair, and $m\,\text{kg}$ the mass of chair and rider.

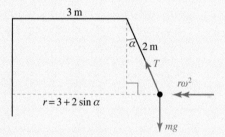

▲ **Figure 4.11**

If the chains make an angle α with the vertical, the motion is in a horizontal circle with radius given by

$$r = 3 + 2\sin\alpha.$$

The magnitude of the acceleration is given by

$$r\omega^2 = (3 + 2\sin\alpha) \times 1.1^2.$$

The acceleration is in a horizontal direction towards the centre of the circle. Using Newton's second law in this direction gives

$$\text{Force} = mr\omega^2$$

$$\Rightarrow T\sin\alpha = m(3 + 2\sin\alpha) \times 1.1^2$$

$$= 1.21m(3 + 2\sin\alpha) \qquad ①$$

→

Answers to exercises are available at www.hoddereducation.com/cambridgeextras

Vertically: $T\cos\alpha - mg = 0$

$\Rightarrow$ $T = \dfrac{mg}{\cos\alpha}$

Substituting for T in equation ①:

$\dfrac{mg}{\cos\alpha}\sin\alpha = 1.21m(3 + 2\sin\alpha)$

$\Rightarrow$ $10\tan\alpha = 3.63 + 2.42\sin\alpha$

 Since m cancels out at this stage, the angle does not depend on the mass of the rider.

This equation cannot be solved directly, but a numerical method will give you the solution 24.9°. You might like to solve the equation yourself or check that this solution does in fact satisfy the equation.

Note

Since the answer does not depend on the mass of the rider and chair, when riders of different masses, or even no riders, are on the equipment all the chains should make the same angle with the vertical.

Banked tracks

ACTIVITY 4.1

Keep away from other people and breakable objects when carrying out this activity.

Place a coin on a piece of stiff A4 card and hold it horizontally at arm's length with the coin near your hand.

▲ Figure 4.12

Turn round slowly so that your hand moves in a horizontal circle. Now gradually speed up. The outcome will probably not surprise you.

What happens, though, if you tilt the card?

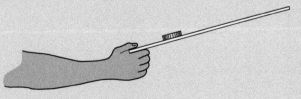

▲ Figure 4.13

You may have noticed that when they curve round bends, most roads are banked so that the edge at the outside of the bend is slightly higher than that at the inside. For the same reason, the outer rail of a railway track is slightly higher than the inner rail when it goes round a bend. On bobsleigh tracks the bends are almost bowl shaped, with a much greater gradient on the outside.

▲ **Figure 4.14**

Figure 4.15 shows a car rounding a bend on a road that is banked so that the cross-section makes an angle α with the horizontal.

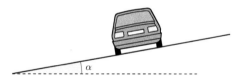

▲ **Figure 4.15**

In modelling such situations, it is usual to treat the bend as part of a horizontal circle with a radius that is large compared to the width of the car. In this case, the radius of the circle is taken to be r metres, and the speed of the car constant at v metres per second. The car is modelled as a particle that has an acceleration of $\frac{v^2}{r} \, \mathrm{m\,s^{-2}}$ in a horizontal direction towards the centre of the circle. The forces and acceleration are shown in Figure 4.16.

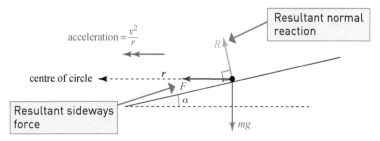

▲ **Figure 4.16**

The direction of the frictional force F will be up or down the slope, depending on whether the car has a tendency to slip sideways towards the inside or outside of the bend.

> ❯ Under what conditions do you think that the car will slip towards (a) the inside or (b) the outside of the bend?

Example 4.5

A car is rounding a bend of radius 100 m that is banked at an angle of 10° to the horizontal. At what speed must the car travel to ensure it has no tendency to slip sideways?

Solution

When there is no tendency to slip there is no frictional force so, in the plane perpendicular to the direction of motion of the car, the forces and acceleration are as shown in Figure 4.17. The only horizontal force is provided by the horizontal component of the normal reaction of the road on the car.

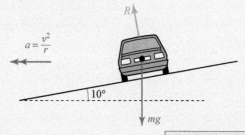

▲ **Figure 4.17**

Vertically, there is no acceleration so there is no resultant force.

> The normal reaction R is resolved into components
> $R \sin 10°$ horizontally ($\leftarrow$)
> $R \cos 10°$ vertically   ($\uparrow$).

$$R \cos 10° - mg = 0$$

$$\Rightarrow \qquad R = \frac{mg}{\cos 10°} \qquad \qquad ①$$

By Newton's second law, in the horizontal direction towards the centre of the circle

$$R \sin 10° = ma = \frac{mv^2}{r}$$

$$= \frac{mv^2}{100}$$

Substituting for R from ①:

$$\left(\frac{mg}{\cos 10°}\right) \sin 10° = \frac{mv^2}{100}$$

$$\Rightarrow \quad v^2 = 100 g \tan 10°$$

> The mass, m, cancels out at this stage, so the answer does not depend on it.

$$\Rightarrow \quad v = 13.3$$

The speed of the car must be about 13.3 m s^{-1}.

There are two important points to notice in Example 4.5.

» The speed is the same whatever the mass of the car.
» The example looks at the situation when the car does not tend to slide, and finds the speed at which this is the case. At this speed the car does not depend on friction to keep it from sliding, and indeed it could travel safely round the bend at this speed even in very icy conditions. However, at other speeds there is a tendency to slide, and friction actually helps the car to follow its intended path.

Safe speeds on a bend

What would happen in Example 4.5 if the car travelled either at a higher or lower speed than $13.3\,\mathrm{m\,s^{-1}}$?

The answer is that there would be a frictional force acting so as to prevent the car from sliding across the road.

There are two possible directions for the frictional force. When the vehicle is stationary or travelling slowly, there is a tendency to slide down the slope and the friction acts up the slope to prevent this. When it is travelling quickly round the bend, the car is more likely to slide up the slope, so the friction acts down the slope.

Fortunately, under most road conditions, the coefficient of friction between tyres and the road is large, typically about 0.8. This means that there is a range of speeds that are safe for negotiating any particular bend.

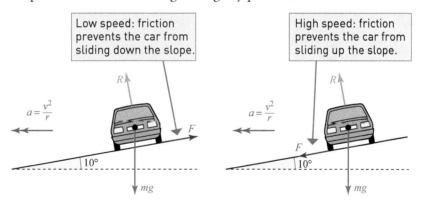

▲ **Figure 4.18**

Answers to exercises are available at www.hoddereducation.com/cambridgeextras

> Using a particle model for the car in Figure 4.18, show that it will not slide up or down the slope provided that

$$\sqrt{rg\frac{(\sin\alpha-\mu\cos\alpha)}{(\cos\alpha+\mu\sin\alpha)}} < v < \sqrt{rg\frac{(\sin\alpha+\mu\cos\alpha)}{(\cos\alpha-\mu\sin\alpha)}}$$

> If $r = 100$ and $\alpha = 10°$ (so that $\tan\alpha = 0.176$) the minimum and maximum safe speeds (in $km\,h^{-1}$) for different values of μ are given in this table.

μ	0	0.1	0.2	0.3	0.4	0.5	0.6	0.7	0.8	0.9	1.0	1.1	1.2
Minimum safe speed	48	31	0	0	0	0	0	0	0	0	0	0	0
Maximum safe speed	48	60	71	81	90	98	106	114	121	129	136	143	150

Would you regard this bend as safe? How, by changing the values of r and α, could you make it safer?

Example 4.6

A bend on a railway track has a radius of $500\,m$ and is to be banked so that a train can negotiate it at $96\,km\,h^{-1}$ without the need for a lateral force between its wheels and the rail. The distance between the rails is $1.43\,m$.

How much higher should the outside rail be than the inside one?

Solution

There is very little friction between the track and the wheels of a train. Any sideways force required is provided by the 'lateral thrust' between the wheels and the rail. The ideal speed for the bend is such that the lateral thrust is zero.

Figure 4.19 shows the forces acting on the train and its acceleration when the track is banked at an angle α to the horizontal.

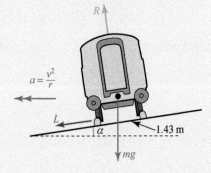

▲ **Figure 4.19**

When there is no lateral thrust, $L = 0$.

Horizontally: $\qquad\qquad R \sin \alpha = \dfrac{mv^2}{r}$ ①

Vertically: $\qquad\qquad R \cos \alpha = mg$ ②

Dividing ① by ② gives $\qquad \tan \alpha = \dfrac{v^2}{rg}$

Using the fact that $96\,\text{km}\,\text{h}^{-1} = 26\frac{2}{3}\,\text{m}\,\text{s}^{-1}$ this becomes

$$\tan \alpha = \frac{32}{225}$$

$$\Rightarrow \quad \alpha = 8.1°$$

> Instead of rounding to three significant figures, it is usual to round angles to one decimal place.

The outside rail should be raised by $1.43 \sin 8.1°$ metres, which is about $20\,\text{cm}$.

Exercise 4B

1 The diagram shows two cars, A and B, travelling at constant speeds in different lanes (radii $24\,\text{m}$ and $20\,\text{m}$) round a circular traffic island. Car A has speed $18\,\text{m}\,\text{s}^{-1}$ and car B has speed $15\,\text{m}\,\text{s}^{-1}$.

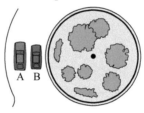

Answer these questions and explain your answers.

(i) Which car has the greater angular speed?

(ii) Is one car overtaking the other?

(iii) Find the magnitude of the acceleration of each car.

(iv) In which direction is the resultant force on each car acting?

2 Two coins are placed on a horizontal turntable. Coin A has mass $15\,\text{g}$ and is placed $5\,\text{cm}$ from the centre; coin B has mass $10\,\text{g}$ and is placed $7.5\,\text{cm}$ from the centre. The coefficient of friction between each coin and the turntable is 0.4.

(i) Describe what happens to the coins when the turntable turns at

(a) $6\,\text{rad}\,\text{s}^{-1}$ (b) $8\,\text{rad}\,\text{s}^{-1}$ (c) $10\,\text{rad}\,\text{s}^{-1}$.

(ii) What would happen if the coins were interchanged?

3 A car is travelling at a steady speed of $15\,\text{m}\,\text{s}^{-1}$ round a roundabout of radius $20\,\text{m}$.

(i) Criticise this false argument:

The car is travelling at a steady speed and so its speed is neither increasing nor decreasing and therefore the car has no acceleration.

(ii) Calculate the magnitude of the acceleration of the car.

(iii) The car has mass 800 kg. Calculate the sideways force on each wheel, assuming it to be the same for all four wheels.

4 A fairground ride has seats at 3 m and at 4.5 m from the centre of rotation. Each rider travels in a horizontal circle. Say whether each of these statements is true, giving your reasons.

(i) Riders in the two positions have the same angular speed at any time.

(ii) Riders in the two positions have the same speed at any time.

(iii) Riders in the two positions have the same magnitude of acceleration at any time.

5 A skater of mass 60 kg follows a circular path of radius 4 m, moving at $2\,\mathrm{m\,s^{-1}}$.

(i) Calculate:

(a) the angular speed of the skater

(b) the magnitude of the acceleration of the skater

(c) the resultant force acting on the skater.

(ii) What modelling assumptions have you made?

6 Two spin dryers, both of which rotate about a vertical axis, have different specifications, as given in this table.

Model	Rate of rotation	Drum diameter
A	600 rpm	60 cm
B	800 rpm	40 cm

State, with reasons, which model you would expect to be the more effective.

7 A satellite of mass M_s is in a circular orbit around the Earth, with a radius of r metres. The force of attraction between the Earth and the satellite is given by

$$F = \frac{GM_eM_s}{r^2}$$

where $G = 6.67 \times 10^{-11}$ in S.I. units. The mass of the Earth M_e is 5.97×10^{24} kg.

(i) Find, in terms of r, expressions for

(a) the speed of the satellite, $v\,\mathrm{m\,s^{-1}}$

(b) the time, T s, it takes to complete one revolution.

(ii) Hence show that, for all satellites, T^2 is proportional to r^3.

A geostationary satellite orbits the Earth so that it is always above the same place on the equator.

(iii) How far is it from the centre of the Earth?

Historical note

The law found in part (ii) was discovered experimentally by Johannes Kepler (1571–1630) to hold true for the planets as they orbit the sun, and is commonly known as Kepler's third law.

8 A rotary lawn mower uses a piece of light nylon string with a small metal sphere on the end to cut the grass. The string is 20 cm in length and the mass of the sphere is 30 g.

(i) Find the tension in the string when the sphere is rotating at 2000 rpm, assuming the string is horizontal.

(ii) Explain why it is reasonable to assume that the string is horizontal.

(iii) Find the speed of the sphere when the tension in the string is 80 N.

9 In this question you should assume that the orbit of the Earth around the Sun is circular, with radius 1.44×10^{11} m, and that the Sun is fixed.

(i) Find the magnitude of the acceleration of the Earth as it orbits the Sun.

The force of attraction between the Earth and the Sun is given by

$$F = \frac{GM_e M_s}{r^2}$$

where M_e is the mass of the Earth, M_s is the mass of the Sun, r the radius of the Earth's orbit and G the universal constant of gravitation (6.67×10^{-11} in S.I. units).

(ii) Calculate the mass of the Sun.

(iii) Comment on the significance of the fact that you cannot calculate the mass of the Earth from the radius of its orbit.

10 Samira ties a model plane of mass 180 g to the end of a piece of string 80 cm long and then swings it round so that the plane travels in a horizontal circle. The plane is not designed to fly and there is no lift force acting on its wings.

(i) Explain why it is not possible for the string to be horizontal.

Samira gives the plane an angular speed of 120 rpm.

(ii) What is the angular speed in radians per second?

(iii) Copy the diagram below and mark in the tension in the string, the weight of the plane and the direction of the acceleration.

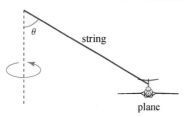

(iv) Write down the horizontal radial equation of motion for the plane and the vertical equilibrium equation in terms of the angle θ.

(v) Show that under these conditions θ has a value between 85° and 86°.

(vi) Find the tension in the string.

Answers to exercises are available at www.hoddereducation.com/cambridgeextras

11 Experiments carried out by the police accident investigation department suggest that a typical value for a coefficient of friction between the tyres of a car and a road surface is 0.8.

(i) Using this information, find the maximum safe speed on a level circular motorway slip road of radius 50 m.

(ii) How much faster could cars travel if the slip road were banked towards the centre of the circle at an angle of 5° to the horizontal?

12 The coefficient of friction between the tyres of a car and the road is 0.8. The mass of the car and its passengers is 800 kg. Model the car as a particle.

(i) Find the maximum frictional force the road can exert on the car and describe what might be happening when this maximum force is acting

(a) at right angles to the line of motion

(b) along the line of motion.

(ii) What is the maximum speed at which the car can travel without skidding round a circular bend of radius 120 m on level ground?

The diagram shows the car, now travelling around a bend of radius 120 m on a road banked at an angle α to the horizontal. The car's speed is such that there is no sideways force (up or down the slope) exerted on its tyres by the road.

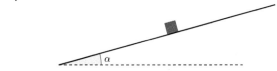

(iii) Draw a diagram showing the weight of the car, the normal reaction of the road on it and the direction of its acceleration.

(iv) Resolve the forces in the horizontal radial and vertical directions and write down the horizontal equation of motion and the vertical equilibrium equation.

(v) Show that $\tan \alpha = \dfrac{v^2}{120g}$ where v is the speed of the car in metres per second.

(vi) On this particular bend, vehicles are expected to travel at 15 m s^{-1}. At what angle, α, should the road be banked?

13 A small sphere S of mass m kg is moving inside a smooth hollow bowl whose axis is vertical and whose sloping side is inclined at 60° to the horizontal. S moves with constant speed in a horizontal circle of radius 0.6 m (see Diagram a). S is in contact with both the plane base and the sloping side of the bowl (see Diagram b).

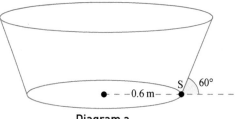

Diagram a

Diagram b

(i) Given that the magnitudes of the forces exerted on S by the base and the sloping side of the bowl are equal, calculate the speed of S.

(ii) Given instead that S is on the point of losing contact with one of the surfaces, find the angular speed of S.

Cambridge International AS & A Level Mathematics
9709 Paper 51 Q3 June 2012

14 Particles P and Q have masses 0.8 kg and 0.4 kg, respectively. P is attached to a fixed point A by a light inextensible string that is inclined at an angle α to the vertical. Q is attached to a fixed point B, which is vertically below A, by a light inextensible string of length 0.3 m. The string BQ is horizontal. P and Q are joined to each other by a light inextensible string that is vertical. The particles rotate in horizontal circles of radius 0.3 m about the axis through A and B with constant angular speed 5 rad s^{-1} (see diagram).

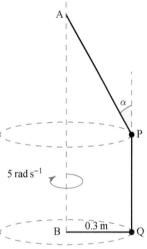

(i) By considering the motion of Q, find the tensions in the strings PQ and BQ.

(ii) Find the tension in the string AP and the value of α.

Cambridge International AS & A Level Mathematics
9709 Paper 53 Q3 November 2010

15 A light inextensible string of length 0.8 m is threaded through a smooth ring and carries a particle at each end. Particle A of mass m kg is at rest at a distance of 0.3 m below the ring. The other particle, B, of mass M kg is rotating in a horizontal circle whose centre is A.

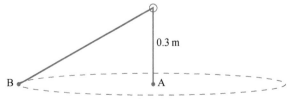

(i) Express M in terms of m.

(ii) Find the angular velocity of B.

Answers to exercises are available at www.hoddereducation.com/cambridgeextras

16 A particle of mass 0.2 kg is moving on the smooth inside surface of a fixed hollow sphere of radius 0.75 m. The particle moves in a horizontal circle whose centre is 0.45 m below the centre of the sphere. (see diagram).

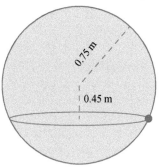

(i)　Show that the force exerted by the sphere on the particle has magnitude $\frac{1}{3}g$.

(ii)　Find the speed of the particle.

(iii)　Find the time taken for the particle to complete one revolution.

17 A particle P of mass 0.25 kg is attached to one end of each of two inextensible strings that are both taut. The other end of the longer string is attached to a fixed point A, and the other end of the shorter string is attached to a fixed point B, which is vertically below A.

String AP is 0.2 m long and string BP is 0.15 m long. P moves in a horizontal circle of radius 0.12 m with constant angular speed $10\,\text{rad}\,\text{s}^{-1}$. Both strings are taut: T_1 is the tension in AP and T_2 is the tension in BP.

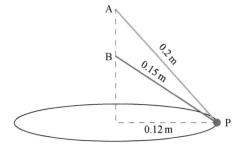

(i)　Resolve vertically to show that $8T_1 + 6T_2 = 2.5g$.

(ii)　Find another equation connecting T_1 and T_2 and hence calculate T_1 and T_2.

4.6 Motion in a vertical circle

Figure 4.20 shows the forces acting on a particle of mass m undergoing free circular motion in a vertical plane. For free motion it is assumed that there is no transverse force.

The weight, *mg*, is resolved into two components,

Radial ($\searrow$):
$mg \cos\theta$

Transverse ($\swarrow$):
$mg \sin\theta$

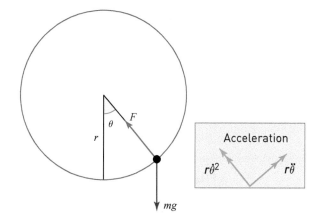

▲ **Figure 4.20**

For circular motion to take place, there must be a resultant force acting on the particle towards the centre of the circle, as you have seen. This is denoted by *F*.

When the circle is vertical, the force of gravity also acts in this plane, and is therefore relevant to the motion. When the particle is in the position shown in Figure 4.20, Newton's second law gives the following equations.

Towards the centre $F - mg \cos\theta = mr\dot{\theta}^2$ ①

Transverse motion $-mg \sin\theta = mr\ddot{\theta}$ ②

The force, *F*, in the first equation might be the tension in a string or the normal reaction from a surface. This force will vary with *θ* and so equation ① is not helpful in describing how *θ* varies with time. The second equation, however, does not involve *F* and may be written as

$$\frac{d^2\theta}{dt^2} = -\frac{g}{r}\sin\theta$$

This differential equation can be solved, using suitable calculus techniques, to obtain an expression for *θ* in terms of *t*. The work is beyond the scope of this book.

Using conservation of energy

A different (and at this stage more profitable) approach is to consider the energy of the particle. Since there is no motion in the radial direction, and no force in the transverse direction, the principle of conservation of mechanical energy can be applied.

Answers to exercises are available at www.hoddereducation.com/cambridgeextras

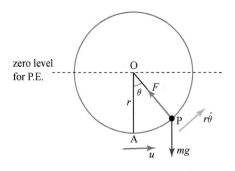

▲ **Figure 4.21**

Take u to be the speed of the particle at A, the lowest point of the circle, and take the zero level of gravitational potential energy to be that through the centre of the circle, O, as shown in Figure 4.21.

The total energy at A is $\frac{1}{2}mu^2 - mgr$

$$\text{(K.E.)} \quad \text{(P.E.)}$$

The total energy at P is $\frac{1}{2}m(r\dot{\theta})^2 - mgr\cos\theta$

$$\text{(K.E.)} \quad \text{(P.E.)}$$

By the principle of conservation of energy

$$\frac{1}{2}m(r\dot{\theta})^2 - mgr\cos\theta = \frac{1}{2}mu^2 - mgr$$

$$\Rightarrow \qquad r\dot{\theta}^2 = \frac{u^2}{r} - 2g(1-\cos\theta)$$

This tells you the angular speed, $\dot{\theta}$, of the particle when OP is at an angle θ to OA.

Examples 4.7 and 4.8 show how conservation of energy may be applied to theoretical models of problems involving motion in a vertical circle.

Example 4.7

A particle of mass 0.03 kg is attached to the end, P, of a light rod, OP, of length 0.5 m that is free to rotate in a vertical circle with centre O. The particle is set in motion, starting at the lowest point of the circle.

The initial speed of the particle is $2\,\mathrm{m\,s}^{-1}$.

(i)   Find the initial kinetic energy of the particle.

(ii) Find an expression for the potential energy gained when the rod has turned through an angle θ.

(iii) Find the value of θ when the particle first comes to rest.

(iv) Find the stress in the rod at this point, stating whether it is a tension or a thrust.

(v)   Repeat parts (i) to (iv) using an initial speed of $4\,\mathrm{m\,s}^{-1}$.

(vi) Why is it possible for the first motion (when $v_0 = 2$) to take place if the rod is replaced by a string, but not the second motion (when $v_0 = 4$)?

Solution

(i) Kinetic energy $= \frac{1}{2}mv^2$

$$= \frac{1}{2} \times 0.03 \times 2^2$$

$$= 0.06$$

The initial kinetic energy is 0.06 J.

(ii) Figure 4.22 shows the position of the particle when the rod has rotated through an angle θ.

It has risen a distance AN where

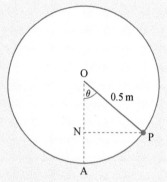

$$AN = OA - ON$$
$$= 0.5 - 0.5\cos\theta$$
$$= 0.5(1 - \cos\theta)$$

The gain in potential energy at P is therefore

▲ Figure 4.22

$$0.03g \times 0.5(1 - \cos\theta) = 0.015g(1 - \cos\theta)\,\text{J}.$$

(iii) When the particle first comes to rest, the kinetic energy is zero, so, by the principle of conservation of energy:

$$0.015g(1 - \cos\theta) = 0.06$$
$$1 - \cos\theta = \frac{0.06}{0.015g}$$
$$= 0.4$$
$$\Rightarrow \qquad \cos\theta = 0.6$$
$$\Rightarrow \qquad \theta = 53.1°$$

(iv) The forces acting on the particle and its acceleration are as shown in Figure 4.23.

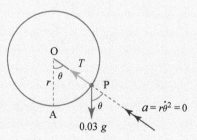

▲ Figure 4.23

Answers to exercises are available at www.hoddereducation.com/cambridgeextras

The component of the acceleration towards the centre of the circle is $r\dot{\theta}^2$, which equals zero when the angular speed is zero. Resolving towards the centre:

$$T - 0.03g\cos\theta = 0 \qquad \textcircled{1}$$
$$T = 0.03 \times 10 \times \cos 53.7° = 0.180$$

Since this is positive, the stress in the rod is a tension. Its magnitude is $0.180\,\text{N}$.

(v)　When the initial speed is $4\,\text{m s}^{-1}$, the initial kinetic energy is
$$\tfrac{1}{2} \times 0.03 \times 4^2 = 0.24\,\text{J}.$$

The gain in potential energy at P, as shown in part (ii),
$$= 0.015g(1 - \cos\theta)\,\text{J}.$$

When the particle first comes to rest, the kinetic energy is zero, so, by the principle of conservation of energy:

$$0.015g(1 - \cos\theta) = 0.24$$
$$\cos\theta = 1 - \frac{0.24}{0.015 \times 10} = -0.6$$
$$\theta = 126.9°$$

Now equation $\textcircled{1}$ gives the tension in the rod as

$$T = 0.03g\cos 126.9° = -0.180\,\text{N}$$

The negative tension means that the stress is in fact a thrust of magnitude $0.180\,\text{N}$.

Figure 4.24 illustrates the forces acting in this position.

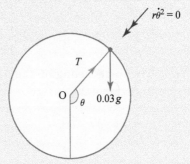

▲ **Figure 4.24**

(vi)　A string cannot exert a thrust, so, although the rod could be replaced by a string in the first case, it would be impossible in the second. In the absence of any radial thrust, the particle would leave its circular path at the point where the tension is zero and before reaching the position where the velocity is zero.

Example 4.8

When a bead is threaded on a wire, it can't fall off.

A bead of mass 0.01 kg is threaded onto a smooth circular wire of radius 0.6 m and is set in motion with a speed of u ms^{-1} at the bottom of the circle. This just enables the bead to reach the top of the wire.

(i)   Find the value of u.

(ii)   What is the direction of the reaction of the wire on the bead when the bead is at the top of the circle?

Solution

(i)   The initial kinetic energy is

$$\frac{1}{2} \times 0.01 u^2 = 0.005 u^2$$

If the bead just reaches the top, the speed there is zero. If this is the case, the kinetic energy at the top will also be zero.

It has then risen a height of $2 \times 0.6 = 1.2$ m, so its gain in potential energy is

$$0.01 g \times 1.2 = 0.012 g$$

By the principle of conservation of energy,

$$\text{loss in K.E.} = \text{gain in P.E.}$$
$$0.005 u^2 = 0.012 g$$
$$u^2 = 2.4 g$$
$$u = \sqrt{2.4 \times 10} = 4.90$$

The initial speed must be 4.90 ms^{-1}.

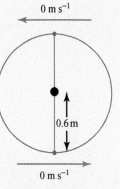

▲ Figure 4.25

(ii)   The reaction of the wire on the bead could be directed either towards the centre of the circle or away from it. The bead has zero angular speed at the top, so the component of its acceleration, and therefore the resultant force towards the centre, is zero. The reaction must be outwards, as shown in Figure 4.26, and equal to $0.01g$ N.

?

Would this motion be possible if the bead were tied to the end of a string instead of being threaded on a wire?

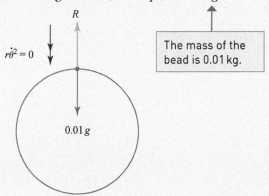

The mass of the bead is 0.01 kg.

▲ Figure 4.26

Answers to exercises are available at <u>www.hoddereducation.com/cambridgeextras</u>

The breakdown of circular motion

> ## ACTIVITY 4.2
>
> Tie a small object to the end of a piece of strong thread and tie the other end loosely (to minimise friction) round a smooth knitting needle (or a smooth rod with a cork on one end).
>
> Hold the pointed end of the needle and make the object move in a vertical circle, as shown in Figure 4.27.
>
>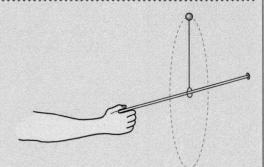
> ▲ **Figure 4.27**
>
> Demonstrate these three types of motion.
>
> (A) The object travels in complete circles.
>
> (B) The object swings like a pendulum.
>
> (C) The object rises above the level of the needle but then fails to complete a full circle.
>
> (i) What would happen if the string broke?
>
> (ii) What would happen if a rod were used rather than a string?
>
> **!** Keep away from other people and breakable objects when carrying out this activity.

The three different types of motion mentioned in Activity 4.2, and the case in which the string breaks, are illustrated in Figure 4.28.

(A) Object oscillates in complete circles

(B) Object oscillates backwards and forwards

(C) Object leaves circle at some point and falls inwards

(D) String breaks and object starts to move away along a tangent

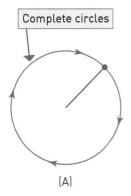

Complete circles

(A)

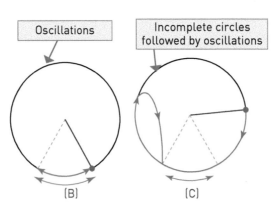

Oscillations

(B)

Incomplete circles followed by oscillations

(C)

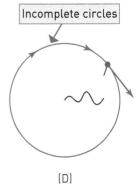

Incomplete circles

(D)

▲ **Figure 4.28**

Modelling the breakdown of circular motion

For what reasons might something depart from motion in a circle? For example, under what conditions will a particle attached to a string and moving in a vertical circle fall out of the circle? Under what conditions will a bicycle travelling over a speed bump with circular cross-section leave the road?

A particle on a string

Figure 4.29 shows a particle P of mass m attached to a string of length r, rotating with angular speed ω in a vertical circle, centre O.

There are two forces acting on the particle, its weight, mg, and the tension, T, in the string. The acceleration of the particle is $r\omega^2$ towards the centre of the circle.

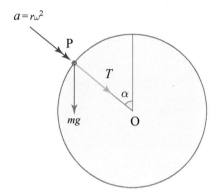

▲ **Figure 4.29**

Applying Newton's second law towards the centre gives

$$T + mg\cos\alpha = mr\omega^2 \qquad\qquad ①$$

where α is the angle shown in the diagram.

While the particle is in circular motion, the string is taut and so $T > 0$. At the instant it starts to leave the circle, the string goes slack and $T = 0$.

Substituting $T = 0$ in ① gives $\quad mg\cos\alpha = mr\omega^2$

$$\Rightarrow \qquad\qquad\qquad \cos\alpha = \frac{r\omega^2}{g}$$

The equation $\cos\alpha = \dfrac{r\omega^2}{g}$ allows you to find the angle α at which the particle leaves the circle, if it does.

The greatest possible value for $\cos\alpha$ is 1, so if $\dfrac{r\omega^2}{g}$ is greater than 1 throughout the motion, the equation has no solution and this means that the particle never leaves the circle. Thus the condition for the particle to stay in circular motion is that $\omega^2 > \dfrac{g}{r}$ throughout.

Answers to exercises are available at www.hoddereducation.com/cambridgeextras

In this example of a particle on a string, ω varies throughout the motion. As you saw earlier, the value of ω at any instant is given by the energy equation, which in this case is

$$\frac{1}{2}mr^2\omega^2 + mgr(1+\cos\alpha) = \frac{1}{2}mu^2$$

where u is the speed of the particle at the lowest point.

A particle moving on the inside of a vertical circle

The same analysis applies to a particle sliding around the inside of a smooth circle like the toy car in Figure 4.30. The only difference is that, in this case, the tension, T, is replaced by the normal reaction, R, of the surface on the particle (see Figure 4.31). When $R = 0$, the particle leaves the surface.

▲ Figure 4.30

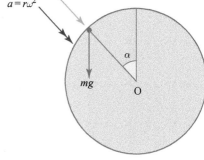

▲ Figure 4.31

A particle moving on the outside of a vertical circle

The forces acting on a particle moving on the outside of a vertical circle, such as a car going over a humpbacked bridge, are the normal reaction, R, acting outwards and the weight of the particle, as shown in Figure 4.32.

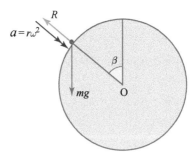

▲ Figure 4.32

Applying Newton's second law towards the centre gives

$$mg\cos\beta - R = mr\omega^2 \qquad ②$$

where β is the angle shown.

If the normal reaction is zero, it means there is no force between the particle and the surface and so the particle is leaving the surface.

Substituting $R = 0$ in ② gives $\quad mg\cos\beta = mr\omega^2$

$$\Rightarrow \qquad \cos\beta = \frac{r\omega^2}{g}$$

> The conditions for the breakdown of circular motion seem to be the same in the cases of a particle on the end of a string and a particle on the outside of a circle.
>
> However, everyday experience tells you that circular motion on the end of a string is only possible if the angular speed is large enough, whereas a particle will only stay on the outside of a circle if the angular speed is small enough.
>
> ❭ How do the conditions $T > 0$ and $R > 0$ explain this difference?

Example 4.9

Determine whether it is possible for a particle, P, of mass m kg to be in the position shown in Figure 4.33 moving round a vertical circle of radius 0.5 m with an angular speed of $4\,\text{rad s}^{-1}$, when it is

(i) sliding on the outside of a smooth surface

(ii) sliding on the inside of a smooth surface

(iii) attached to the end of a string OP

(iv) threaded on a smooth vertical ring.

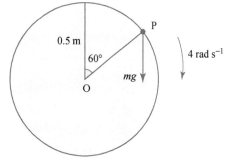

▲ **Figure 4.33**

Solution

(i) On the outside of a smooth surface:

The normal reaction R N of the surface on the particle must be acting outwards, so Newton's second law towards the centre gives

$mg\cos 60° - R = m \times 0.5 \times 4^2$

$$\Rightarrow \qquad R = mg\cos 60° - 8m$$

$$= -3\,m$$

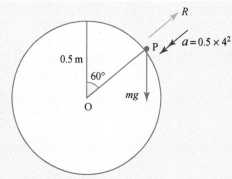

▲ **Figure 4.34**

Whatever the mass, m, this negative value of R is impossible, so the motion is impossible. The particle will already have left the surface.

(ii) On the inside of a smooth surface:

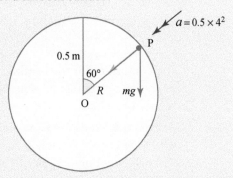

▲ Figure 4.35

The normal reaction of the surface on the particle will now be acting towards the centre and so

$$R + mg\cos 60° = m \times 0.5 \times 4^2$$

$$\Rightarrow \qquad R = +3\,m$$

This is possible.

(iii) Attached to the end of a string:

This situation is like that in part (ii) since the tension acts towards the centre, so the motion is possible.

(iv) Threaded on a smooth ring:

If the particle is threaded on a ring the normal reaction can act inwards or outwards so the motion can take place whatever the angular speed. This is also the case when a particle is attached to the centre by a light rod. The rod will exert a tension or a thrust as required.

? Which of the situations in Example 4.9 are possible when the angular speed is $3\,\text{m}\,\text{s}^{-1}$?

Example 4.10

Eddie, a skier of mass $m\,$kg, is skiing down a hillside when he reaches a smooth hump in the form of an arc AB of a circle with centre O and radius 8 m, as shown in Figure 4.36. O, A and B lie in a vertical plane and OA and OB make angles of 20° and 40° with the vertical, respectively. Eddie's speed at A is $7\,\text{m}\,\text{s}^{-1}$. Determine whether Eddie will lose contact with the ground before reaching the point B.

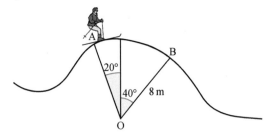

▲ Figure 4.36

Solution

▲ **Figure 4.37**

Note

S is a general
point on the
surface of the arc
of the circle.

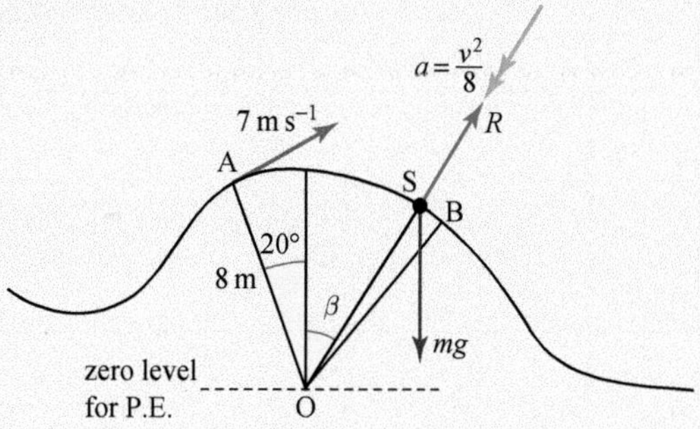

Taking the zero level for potential energy to be a horizontal line through O, the initial energy at A is

$$\frac{1}{2}m \times 7^2 + mg \times 8\cos 20°$$

The energy at point S is

$$\frac{1}{2}mv^2 + mg \times 8\cos\beta$$

By the principle of conservation of energy these are equal.

$$\frac{1}{2}mv^2 + mg \times 8\cos\beta = \frac{1}{2}m \times 7^2 + mg \times 8\cos 20°$$

$$\Rightarrow \quad v^2 + 16g\cos\beta = 49 + 150.3\ldots$$

$$\Rightarrow \qquad\qquad v^2 = 199.3\ldots - 16g\cos\beta \qquad\qquad ①$$

Using Newton's second law towards the centre of the circle

$$mg\cos\beta - R = m\frac{v^2}{8}$$

$$\Rightarrow \qquad R - m\left(g\cos\beta - \frac{v^2}{8}\right)$$

If Eddie leaves the circle at point S, then $R = 0$

$$\Rightarrow \quad v^2 = 8g\cos\beta$$

Substituting in ①

$$8g\cos\beta = 199.3\ldots - 16g\cos\beta$$

$$\Rightarrow \quad 24g\cos\beta = 199.3\ldots$$

$$\Rightarrow \qquad \cos\beta = \frac{199.3\ldots}{24 \times 10} = 0.830\ldots$$

$$\Rightarrow \qquad \beta = 33.8°$$

This gives $\beta = 33.8°$, which is less than $40°$, so Eddie will lose contact with the ground before he reaches the point B.

1   The diagrams show two particles of mass m kg moving in vertical circles. Their angular speeds and positions are as shown.

(A)

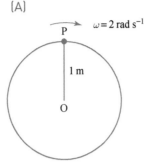

(B)

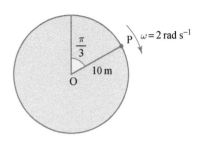

(i)   Write down the components of their acceleration along PO.

(ii) By considering the forces acting on the particle, determine, in each case, whether it is possible for it to be moving with this speed in this position when it is:

(a) sliding on the outside of a smooth surface

(b) sliding on the inside of a smooth surface

(c) attached to the end of a string OP

(d) threaded on a smooth ring in a vertical plane.

2   Each diagram shows a particle of mass m that is constrained to move in a vertical circle. Initially, it is in the position shown and moving with the given speed.

(A)   (B)   (C)

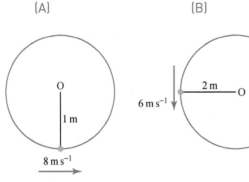
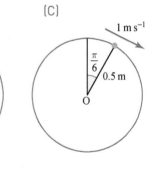

In each case:

(i)   Using the horizontal through the centre of the circle as the zero level for potential energy, write down the total initial mechanical energy of the particle.

(ii) Decide whether the particle will make complete revolutions or whether it will come to rest below the highest point. If it makes complete revolutions, determine the speed at the top. Otherwise, find the height above the centre when it comes to rest.

(iii) Assuming that the particle is moving under the action of its weight and a radial force only, find the magnitude and direction of the radial force

(a) initially   (b)   when it reaches the top or comes to rest.

3 A smooth hemispherical bowl of radius r, with lowest point A, is fixed with its rim uppermost and horizontal. A particle of mass m is projected along the inner surface of the bowl with a speed $\sqrt{gr}$ towards A, from a point at a vertical height $\frac{1}{2}r$ above A, so that its motion is in a vertical plane through A.

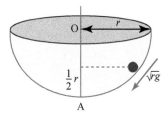

(i) Show that the particle will just reach the top of the bowl.

(ii) Find the reaction between the particle and the bowl when it is at a height $\frac{1}{3}r$ above A.

4 A bead B of mass m is threaded onto a smooth circular wire, fixed in a vertical plane, with centre O and radius a. When the bead is at the lowest point of the wire, it is projected horizontally with velocity u and, in the subsequent motion, B reaches a maximum vertical height of $\frac{1}{2}a$ above O.

(i) Show that $u^2 = 3ga$.

(ii) If θ is the angle that OB makes with the downward vertical, prove that the reaction R of B on the wire is given by

$$R = mg(1 + 3\cos\theta).$$

CP

5 The diagram shows a model car track. You may assume that all parts of the track lie in the same vertical plane.

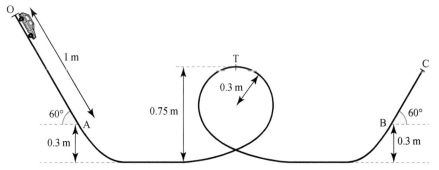

Between A and B the track is symmetrical about a vertical line through T; its length from A to T is 2.1 m. OA and BC are straight and the top of the loop is an arc of a circle of radius 0.3 m. For a car of mass m kg there is a frictional resistance of $0.06mg$ N.

The car starts from rest at the point O.

(i) Find the work done against friction between O and T.

(ii) Use the work–energy principle to show that the kinetic energy at T is $0.230mg$ J.

Answers to exercises are available at www.hoddereducation.com/cambridgeextras

(iii) By considering circular motion at T, show that the car will move right round the loop in contact with the track.

(iv) The car stops at C before returning. Find the length BC.

(v) Will the car reach T on the return journey?

6 A particle of mass m hangs by a string of length a from a fixed point. The particle is given a horizontal velocity of $\sqrt{\frac{7}{2}ga}$.

(i) Show that the string will be about to become slack when it makes an angle of 60° with the upward vertical.

(ii) Find the tension in the string when it makes an angle of 60° with the downward vertical.

7 A metal sphere of mass 0.5 kg is moving in a vertical circle of radius 0.8 m at the end of a light, inelastic string. At the top of the circle the sphere has speed 3 m s⁻¹.

(i) Calculate the gravitational potential energy lost by the sphere when it reaches the bottom of the circle and hence calculate its speed at this point.

(ii) Find an expression for the speed of the sphere when the string makes an angle, θ, with the upward vertical.

(iii) Find the tension in the string when the sphere is

(a) at the top of the circle

(b) at the bottom of the circle.

(iv) Draw a diagram showing the forces acting on the sphere when the string makes an angle, θ, with the upward vertical. Find expressions for the tension in the string and the transverse component of the sphere's acceleration at this instant.

8 A glider is travelling horizontally until the pilot executes a loop-the-loop manoeuvre, as shown in the diagram. The loop may be modelled as a vertical circle. The glider is initially at a height of 700 m, travelling at 30 m s⁻¹. The bottom of the loop is at a height of 400 m and the radius of the loop is 100 m.

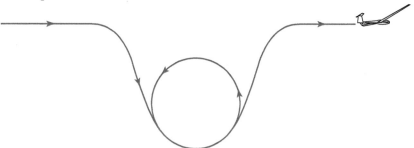

Assuming that the mechanical energy is conserved, calculate:

(i) the speed of the glider at the lowest and highest points of the loop

(ii) the magnitude of the acceleration of the glider at the lowest and highest points of the loop.

The mass of the pilot is 70 kg.

(iii) Draw diagrams to show the reaction forces acting on him at the lowest and highest points of the loop and state their magnitudes.

(iv) What would happen if he attempted a loop of radius 150 m, starting from the same lowest point?

(v) What is the maximum radius for a successful loop from this point?

9 The diagram shows a ride at an amusement park. The loop is, to a good approximation, a circle of radius 8 m, in a vertical plane.

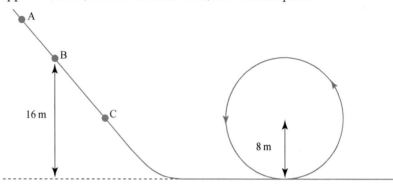

In answering these questions, you should assume that no energy is lost to forces such as friction and air resistance, and that the car starts from rest.

(i) Explain why a car that starts just above point B, 16 m above ground level, will not complete the loop.

(ii) For a car to complete the loop successfully, it must start at or above point A. What is the height of A?

(iii) On 'Kiddie days' the organisers start the car at point C. Describe what happens to the car and state the maximum height of point C for it to be safe.

10 The diagram illustrates an old road bridge over a river. The road surface follows an approximately circular arc with radius 15 m.

A car is being driven across the bridge and you should model it as a particle.

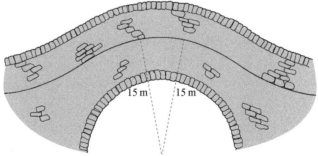

(i) Calculate the greatest constant speed at which it is possible to drive the car across the bridge without it leaving the road.

(ii) Comment on the fact that the bridge is old.

(iii) How is it possible to improve the design of the bridge?

Answers to exercises are available at www.hoddereducation.com/cambridgeextras

11 A car of mass 300 kg travels along the track of a roller coaster. In one section of the track, the car travels around the inside of a vertical circle of radius 7.5 m, as shown in the diagram.

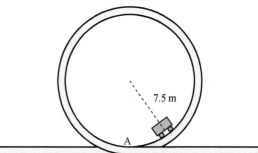

The car is attached to the track so that it can move freely along it, but cannot leave the circular path. While the car is travelling around the circle, there is no driving force present. Friction and air resistance may be neglected.

The car enters the circle at its lowest point, A, with a speed of 20 ms^{-1}.

(i) Show that the speed of the car at the highest point is 10 ms^{-1}.

(ii) Find the radial acceleration at the highest point. Hence calculate the force that the track exerts on the car at this point. Could the roller coaster operate if the car was not attached to the track?

Explain your answer briefly.

(iii) Calculate the radial and tangential components of acceleration when the car has travelled through 120° round the circle from A. Find the magnitude of the resultant acceleration and the direction it makes with the vertical.

12 A particle of mass m is attached to one end, A, of a light inelastic string of length l. The other end of the string, B, is attached to a ceiling so that the particle is free to swing in a vertical plane. The angle between the string and the downward vertical is θ radians. You may assume that the air resistance on the particle is negligible.

Initially, $\theta = \frac{\pi}{3}$ and the particle is released from rest.

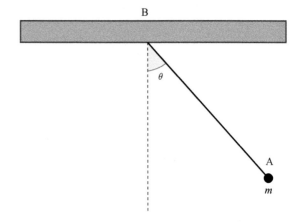

(i) Show that the potential energy lost by the particle since leaving its initial position is $\frac{mgl}{2}(2\cos\theta - 1)$. Hence find an expression for v^2, where v is the linear speed of the particle, in terms of l, g and θ.

(ii) Show that the tension in the string at any point of the motion is $mg(3\cos\theta - 1)$.

(iii) Find the greatest tension in the string. What is the position of the particle when the tension in the string is greatest?

KEY POINTS

1   Position, velocity and acceleration of a particle moving on a circle of radius r.

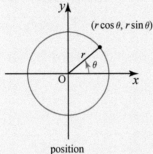 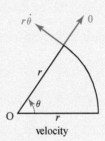 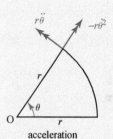

position	velocity	acceleration

position $(r\cos\theta, r\sin\theta)$

velocity   transverse component: $v = r\dot{\theta} = r\omega$
radial component: 0
where $\dot{\theta}$ or ω is the angular velocity of the particle.

acceleration   transverse component: $r\ddot{\theta} = r\dot{\omega}$
radial component: $-r\dot{\theta}^2 = -r\omega^2 = -\dfrac{v^2}{r}$
where $\ddot{\theta}$ or $\dot{\omega}$ is the angular acceleration of the particle.

2   By Newton's second law, the forces acting on a particle of mass m in circular motion are equal to

transverse component:

$mr\dot{\omega} = mr\ddot{\theta}$

radial component:

$mr\dot{\theta}^2 = mr\omega^2 = m\dfrac{v^2}{r}$ towards the centre

$-mr\dot{\theta}^2 = -mr\omega^2 = -m\dfrac{v^2}{r}$ away from the centre

3   Problems involving free motion in a vertical circle can be solved using the conservation of energy principle

P. E. + K.E. = constant

in conjunction with Newton's second law.

4   Circular motion breaks down when the available force towards the centre is $< mr\omega^2$.

Answers to exercises are available at www.hoddereducation.com/cambridgeextras

LEARNING OUTCOMES

Now that you have finished this chapter, you should be able to

- understand the language and notation associated with circular motion
- identify the forces acting on a body in circular motion
- calculate acceleration towards the centre of circular motion
- model situations involving circular motion with uniform speed in a horizontal plane
- model situations involving circular motion with non-uniform speed in a horizontal plane
- model situations involving motion in a vertical circle
- identify the conditions under which a particle departs from circular motion.

5 Hooke's law

The photographs show people bungee jumping.

Bungee jumping is a dangerous sport that originated in the South Sea Islands, where creepers were used instead of ropes. In the more modern version, a person jumps off a high bridge or crane to which they are attached by an elastic rope around their ankles, or with a harness.

> ❯ If somebody bungee jumping from a bridge wants the excitement of *just* reaching the surface of the water below, how would you calculate the length of rope required?

The answer to this question clearly depends on the height of the bridge, the mass of the person jumping and the elasticity of the rope. All ropes are elastic to some extent, but it would be extremely dangerous to use an ordinary rope for this sport because the impulse necessary to stop somebody falling would involve a very large tension acting in the rope for a short time and this would provide too great a shock to the system. A bungee is a strong elastic rope, similar to those used to secure loads on cycles, cars or lorries, with the essential property for this sport that it allows the impulse to act over a much longer time so that the rope exerts a smaller force on the jumper.

Generally, in mechanics, the word **string** is used to represent such things as ropes that can be in tension but not in compression. In this chapter, you will be studying some of the properties of elastic strings and springs and will return to the problem of the bungee jumper as a final investigation.

In contrast, a **spring** can be compressed as well as stretched. So a spring can be in compression or in tension. In this book, it is assumed that springs are open coiled.

5.1 Strings and springs

So far, in situations involving strings, it has been assumed that they do not stretch when they are under tension. Such strings are called **inextensible**. For some materials this is a good assumption, but for others the length of the string increases significantly under tension. Strings and springs that stretch are said to be **elastic**.

The length of a string or spring when there is no force applied to it is called its **natural length**. If it is stretched, the increase in length is called its **extension**, and if a spring is compressed, it is said to have a **negative extension** or **compression**.

When stretched, a spring exerts an inward force, or **tension,** on whatever is attached to its ends (Figure 5.1(B)). When compressed, it exerts an outward force, or **thrust,** on its ends (Figure 5.1 (C)). An elastic string exerts a tension when stretched, but exerts no force when slack.

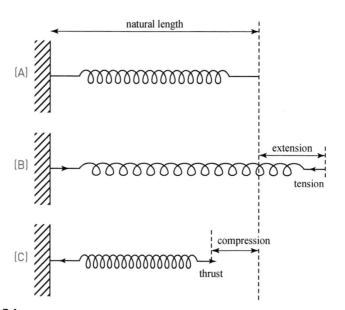

▲ **Figure 5.1**

You will need some elastic strings, some open coiled springs, some weights and a support stand. Set up the apparatus as shown in Figure 5.2.

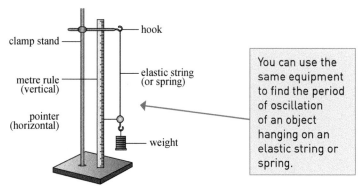

You can use the same equipment to find the period of oscillation of an object hanging on an elastic string or spring.

▲ **Figure 5.2**

Before doing any experiments, predict the answers to these questions.

1   How are the extension of a string and the weight of an object hung from it related?

2   If a string of the same material but twice the natural length has the same weight attached, how does the extension change?

3   Does the string return to its original length when unloaded

   (i) if the weight of the object is small?

   (ii) if the weight of the object is large?

Now, for each string, plot a graph of tension, which is the weight of the object (vertical axis) against the extension (horizontal axis) to help you to answer the questions.

Design and carry out an experiment to investigate the relationship between the thrust in an open coiled spring and the decrease in its length.

From your experiments, you should have made these observations.

▸▸ Each string or spring returns to its original length once the object is removed, up to a certain limit.

Strings or springs that exhibit this linear behaviour are said to be **perfectly elastic**.

▸▸ The graph of tension or thrust against extension for each string or spring is a straight line for all or part of the data.

▸▸ The gradient of the linear part of the graph is roughly halved when the string is doubled in length.

▸▸ If you keep increasing the weight, the string or spring may stop stretching or may stretch without returning to its original length. In this case, the graph is no longer a straight line: the material has passed its **elastic limit**.

▸▸ During your experiment using an open coiled spring you may have found it necessary to prevent the spring from buckling. You may also have found that there comes a point when the coils are completely closed and a further decrease in length is impossible.

Answers to exercises are available at www.hoddereducation.com/cambridgeextras

5.2 Hooke's law

In 1660, Robert Hooke discovered a *rule or law of nature in every springing body* which, for small extensions relative to the length of the string or spring, can be stated as follows:

> **The tension in an elastic spring or string is proportional to the extension. If a spring is compressed, the thrust is proportional to the decrease in length of the spring.**

When a string or spring is described as elastic, it means that it is reasonable to apply the modelling assumption that it obeys Hooke's law. A further assumption, that it is light (has zero mass), is usual and is made in this book.

There are three forms in which Hooke's law is commonly expressed for a string. Which one you use depends on the extent to which you are interested in the string itself rather than just its overall properties. Denoting the natural length of the string by l_0 and its area of cross-section by A, the different forms are as follows.

» $T = \dfrac{EA}{l_0} x$ In this form, E is called **Young's modulus** and is a property of the material from which the string is made. This form is commonly used in physics and engineering, subjects in which properties of materials are studied. It is rarely used in mathematics. The S.I. unit for Young's modulus is $N\,m^{-2}$.

» $T = \dfrac{\lambda}{l_0} x$ The constant λ is called the **modulus of elasticity** of the string and will be the same for any string of a given cross-section made from the same material. Many situations require knowledge of the natural length of a string and this form may well be the most appropriate in such cases. The S.I. unit for the modulus of elasticity is N.

» $T = kx$ In this simplest form, k is called the **stiffness** of the string. It is a property of the string as a whole. You may choose to use this form if neither the natural length nor the cross-sectional area of the string is relevant to the situation. The S.I. unit for stiffness is $N\,m^{-1}$.

Notice that $k = \dfrac{\lambda}{l_0} = \dfrac{EA}{l_0}$

In this book, only the form using the modulus of elasticity is used, and this can be applied to springs as well as strings.

| Example 5.1 | A light elastic string of natural length 0.7 m and modulus of elasticity 50 N has one end fixed and a particle of mass 1.4 kg attached to the other. The system hangs vertically in equilibrium. Find the extension of the string. |

Solution

The forces acting on the particle are the tension, T N, upwards and the weight, $1.4g$ N downwards.

Since the particle is in equilibrium
$$T = 1.4g$$

Using Hooke's law: $\quad T = \dfrac{\lambda}{l_0}x$

$$\Rightarrow \qquad 1.4g = \dfrac{50}{0.7}x$$

$$\Rightarrow \qquad x = \dfrac{0.7 \times 1.4g}{50}$$

$$= 0.196$$

The extension in the string is 0.196 m.

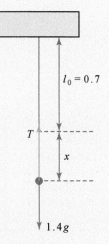

▲ Figure 5.3

| Example 5.2 | The mechanism of a set of kitchen scales consists of a light scale pan supported on a spring of natural length 20 mm. When measuring 1.5 kg of flour, the spring is compressed by 7.5 mm. Find |

(i)   the modulus of elasticity of the spring

(ii)   the mass of the heaviest object that can be measured if it is impossible to compress the spring by more than 16 mm.

Solution

(i)   The forces on the scale pan with its load of flour are the weight, $1.5g$ N, downwards, and the thrust of the spring, T N, upwards.

▲ Figure 5.4

Answers to exercises are available at <u>www.hoddereducation.com/cambridgeextras</u>

Since it is in equilibrium

$$T = 1.5g$$

Applying Hooke's law with modulus of elasticity λ N:

$$T = \frac{\lambda}{0.02} \times 0.0075$$

$$\Rightarrow \qquad 1.5g = 0.375\lambda$$

$$\Rightarrow \qquad \lambda = 40$$

The modulus of elasticity of the spring is 40 N.

(ii)   Let the mass of the heaviest object be M kg, so the maximum thrust is Mg N. Then Hooke's law for a compression of 16 mm gives:

$$Mg = \frac{40}{0.02} \times 0.016$$

$$\Rightarrow \qquad M = 3.2$$

The mass of the heaviest object that can be measured is 3.2 kg.

Note

These scales would probably be calibrated to a maximum of 3 kg.

Exercise 5A

1   In each of these diagrams an object is suspended by a light elastic string. The top of the string is attached to a fixed ceiling. The object is in equilibrium.

In each diagram, information is provided about the mass of the object and two out of the natural length, l_0 m, the length, l m, and the extension, x m.

In each case, find

    (a)   the tension in the string

    (b)   the modulus of elasticity of the string.

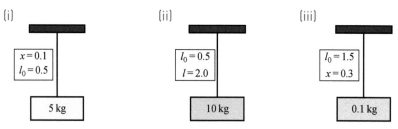

(i)
$x = 0.1$
$l_0 = 0.5$
5 kg

(ii)
$l_0 = 0.5$
$l = 2.0$
10 kg

(iii)
$l_0 = 1.5$
$x = 0.3$
0.1 kg

2   An elastic string has natural length 20 cm. The string is fixed at one end. When a force of 20 N is applied to the other end, the string doubles in length.

    (i)   Find the modulus of elasticity.

    (ii)   Another elastic string also has natural length 20 cm. When a force of 20 N is applied to each end, the string doubles in length.

        Find the modulus of elasticity.

    (iii)   Explain the connection between the answers to parts (i) and (ii).

3   A light spring has modulus of elasticity 0.4 N and natural length 50 cm. One end is attached to a ceiling, the other to a particle of weight 0.03 N that hangs in equilibrium below the ceiling.

(i)   Find the tension in the spring.

(ii) Find the extension of the spring.

The particle is removed and replaced with one of weight w N. When this hangs in equilibrium, the spring has length 60 cm.

(iii) What is the value of w?

4   An object of mass 0.5 kg is attached to an elastic string with natural length 1.2 m and causes an extension of 8 cm when the system hangs vertically in equilibrium.

(i)   What is the tension in the string?

(ii) What is the modulus of elasticity of the string?

(iii) What is the mass of an object that causes an extension of 10 cm?

5   The diagram shows a spring of natural length 60 cm that is being compressed under the weight of a block of mass m kg. Smooth supports constrain the block to move only in the vertical direction.

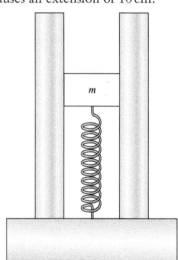

The modulus of elasticity of the spring is 180 N. The system is in equilibrium and the length of the spring is 50 cm. Find

(i)   the thrust in the spring

(ii) the value of m.

More blocks are piled on.

(iii) Describe the situation when there are seven blocks in total, all identical to the first one.

6   An open coiled spring has natural length 30 cm and modulus of elasticity 80 N.

The spring is fully compressed when its length is 15 cm.

If the spring is extended to twice its natural length it passes its elastic limit and can no longer return to its natural length.

Find the limits of the applied force for which Hooke's law may be used as a model for this spring.

7   A small sphere, A, of mass m kg moves in a circle with centre B on a smooth horizontal table. A is joined to a smoothly rotating vertical axle at B by an elastic string of natural length a m and modulus of elasticity λ N and has constant angular speed ω rad s^{-1}. Find an expression for the radius of the circle in terms of m, a, λ and ω.

Answers to exercises are available at www.hoddereducation.com/cambridgeextras

5.3 Using Hooke's law with more than one spring or string

Hooke's law allows you to investigate situations involving two or more springs or strings in various configurations.

Example 5.3

A particle of mass $0.4\,kg$ is attached to the midpoint of a light elastic string of natural length $1\,m$ and modulus of elasticity $\lambda\,N$. The string is then stretched between a point A at the top of a doorway and a point B that is on the floor, $2\,m$ vertically below A.

(i)   Find, in terms of λ, the extensions of the two parts of the string.

(ii) Calculate their values in the case when $\lambda = 10$.

(iii) Find the minimum value of λ that will ensure that the lower half of the string is not slack.

Solution

For a question like this, it is helpful to draw two diagrams, one showing the relevant natural lengths and extensions, and the other showing the forces acting on the particle.

Since the force of gravity acts downwards on the particle, its equilibrium position will be below the midpoint of AB. This is also shown in Figure 5.5.

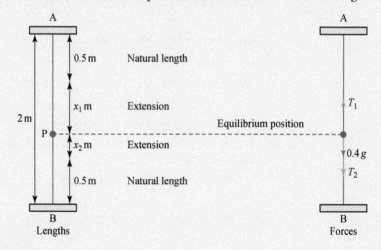

▲ **Figure 5.5**

(i) The particle is in equilibrium, so the resultant vertical force acting on it is zero.

Therefore $\quad\quad T_1 = T_2 + 0.4g \quad\quad\quad\quad\quad$ ①

Hooke's law can be applied to each part of the string.

For AP: $\quad\quad\quad T_1 = \dfrac{\lambda}{0.5}x_1 \quad\quad\quad\quad$ ②

For BP: $\quad\quad\quad T_2 = \dfrac{\lambda}{0.5}x_2 \quad\quad\quad\quad$ ③

Substituting these expressions in equation ① gives:

$$\frac{\lambda}{0.5}x_1 = \frac{\lambda}{0.5}x_2 + 0.4g$$

$$\Rightarrow \quad \lambda(x_1 - x_2) = 0.5 \times 0.4g$$

$$\Rightarrow \quad x_1 - x_2 = 0.2\frac{g}{\lambda} \quad\quad\quad\quad ④$$

But from the first diagram it can be seen that:

$$x_1 + x_2 = 1 \quad\quad\quad\quad ⑤$$

Adding ④ and ⑤ gives:

$$2x_1 = 1 + 0.2\frac{g}{\lambda}$$

$$\Rightarrow \quad x_1 = 0.5 + 0.1\frac{g}{\lambda}$$

Similarly subtracting ④ from ⑤ gives:

$$x_2 = 0.5 - 0.1\frac{g}{\lambda} \quad\quad\quad\quad ⑥$$

(ii) Since $\lambda = 10$, the extensions are $0.6\,\text{m}$ and $0.4\,\text{m}$.

(iii) The lower part of the string will not become slack providing $x_2 > 0$. It follows from equation ⑥ that:

$$0.5 - 0.1\frac{g}{\lambda} > 0$$

$$\Rightarrow \quad\quad\quad 0.5 > 0.1\frac{g}{\lambda}$$

$$\Rightarrow \quad\quad\quad \lambda > 0.2g$$

The minimum value of λ for which the lower part of the string is not slack is $2\,\text{N}$, and in this case BP has zero tension.

Exercise 5B

1   The diagram shows a uniform plank of weight 120 N symmetrically suspended in equilibrium by two identical elastic strings, each of natural length 0.8 m and modulus of elasticity 1200 N.

(i)   Find

(a)   the tension in each string

(b)   the extension of each string.

The two strings are replaced by a single string, also of natural length 0.8 m, attached to the middle of the plank. The plank is in the same position.

(ii)   Find the modulus of elasticity of this string and comment on its relationship to that of the original strings.

2   The manufacturer of a sports car specifies the coil spring for the front suspension as a spring of 10 coils with a natural length 0.3 m and a compression 0.1 m when under a load of 4000 N.

(i)   Calculate the modulus of elasticity of the spring.

(ii)   If the spring were cut into two equal parts, what would be the modulus of elasticity of each part?

The weight of the car is 8000 N and half of this weight is taken by two such 10-coil front springs so that each bears a load of 2000 N.

(iii)   Find the compression of each spring.

(iv) Two people, each of weight 800 N, get into the front of the car. How much further are the springs compressed? (Assume that their weight is carried equally by the front springs.)

PS 3 The coach of an impoverished rugby club decides to construct a scrummaging machine as illustrated in the diagram below. It is to consist of a vertical board, supported in horizontal runners at the top and bottom of each end. The board is held away from the wall by two springs, as shown, and the players push the board with their shoulders, against the thrust of the springs.

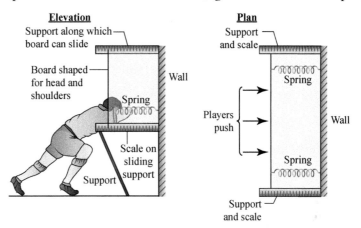

The coach has one spring of length 1.4 m and modulus of elasticity 7000 N, which he cuts into two pieces of equal length.

(i) Find the modulus of elasticity of each of the half-length springs.

(ii) On one occasion, the coach observes that the players compress both springs by 20 cm. What total force do they produce in the forward direction?

PS 4 The diagram shows the rear view of a load of weight 300 N in the back of a pickup truck of width 2 m.

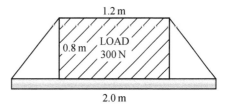

The load is 1.2 m wide, 0.8 m high and is situated centrally on the truck. The coefficient of friction between the load and the truck is 0.4. The load is held down by an elastic rope of natural length 2 m and modulus of elasticity 400 N, which may be assumed to pass smoothly over the corners and across the top of the load. The rope is secured at the edges of the truck platform.

(i) Find

(a) the tension in the rope

(b) the normal reaction of the truck on the load

(ii) the percentage by which the maximum possible frictional force is increased by using the rope

Answers to exercises are available at www.hoddereducation.com/cambridgeextras

(iii) the shortest stopping distance for which the load does not slide, given that the truck is travelling at 30 m s⁻¹ initially.
(Assume constant deceleration and use 10 for g.)

5   The diagram shows two light springs, AP and BP, connected at P.
The ends A and B are secured firmly and the system is in equilibrium.

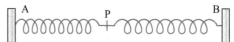

The spring AP has natural length 1 m and modulus of elasticity 16 N.

The spring BP has natural length 1.2 m and modulus of elasticity 30 N.

The distance AB is 2.5 m and the extension of the string AP is x m.

(i)   Write down an expression, in terms of x, for the extension of the spring BP.

(ii) Find expressions, in terms of x, for the tensions in both springs.

(iii) Find the value of x.

6   The diagram shows two light springs, CQ and DQ, connected to a particle, Q, of weight 20 N. The ends C and D are secured firmly and the system is in equilibrium, lying in a vertical line.

The spring CQ has natural length 0.8 m and modulus of elasticity 16 N; DQ has natural length 1.2 m and modulus of elasticity 36 N. The distance CD is 3 m and QD is h m.

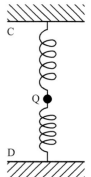

(i)   Write down expressions, in terms of h, for the extensions of the two springs.

(ii) Find expressions, in terms of h, for the tensions in the two springs.

(iii) Use these results to find the value of h.

(iv) Find the forces the system exerts at C and at D.

7   The diagram shows a block of wood of mass m lying on a plane inclined at an angle α to the horizontal. The block is attached to a fixed peg by means of a light elastic string of natural length l_0 and modulus of elasticity λ; the string lies parallel to the line of greatest slope. The block is in equilibrium.

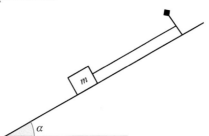

Find the extension of the string in these cases.

(i) The plane is smooth.

(ii) The coefficient of friction between the plane and the block is μ $(\mu \neq 0)$ and the block is about to slide

   (a) up the plane

   (b) down the plane.

8 A particle A and a block B are attached to opposite ends of a light elastic string of natural length 2 m and modulus of elasticity 6 N. B is at rest on a rough horizontal table. The string passes over a small smooth pulley P at the edge of the table, with the part BP of the string horizontal and of length 1.2 m.

The frictional force acting on B is 1.5 N and the system is in equilibrium. Find the distance PA.

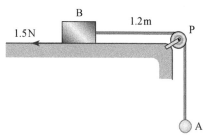

Cambridge International AS & A Level Mathematics
9709 Paper 5 Q1 June 2008

9 A strong elastic band of natural length 1 m and of modulus of elasticity 12 N is stretched round two pegs, P and Q, which are in a horizontal line a distance of 1 m apart. A bag of mass 1.5 kg is hooked onto the band at H and hangs in equilibrium so that PH and QH make angles of θ with the horizontal. Make the modelling assumptions that the elastic band is light and runs smoothly over the pegs.

(i) Use Hooke's law to show that the tension in the band is $12 \sec \theta$.

(ii) Find the depth of the hook below the horizontal line PQ.

(iii) Is the modelling in this question realistic? Justify your answer.

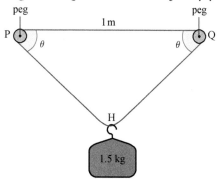

10 A light elastic string has natural length 0.6 m and modulus of elasticity λ N.

The ends of the string are attached to fixed points A and B, which are at the same horizontal level and 0.63 m apart. A particle P of mass 0.064 kg is attached to the midpoint of the string and hangs in equilibrium at a point 0.08 m below AB (see diagram).

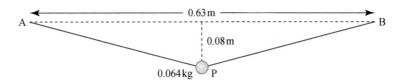

Find

(i)   the tension in the string

(ii)   the value of λ.

Cambridge International AS & A Level Mathematics
9709 Paper 5 Q1 June 2006

11 A spring AB of natural length 0.5 m and modulus of elasticity 25 N is fixed at A.

The other end is joined to another spring BC of natural length 0.8 m and modulus of elasticity 40 N.

A weight W N is attached at C and the system hangs vertically in equilibrium so that AC = 2 m.

(i)   Find the extension of the two springs.

(ii)   Find W.

CP **12** A light elastic string of natural length l_0 is hung at one end from a fixed point. When a particle of mass m is hung from the other end, the string extends a distance d. Show that the modulus of elasticity of the string is $\frac{mgl_0}{d}$.

The particle is removed and attached to the midpoint of the string. The ends of the string are now tied to two points, A and B, where B is vertically below A and AB > l_0. In the equilibrium position, the lower part of the string remains taut. Show that the displacement of the particle from the midpoint of AB is $\frac{d}{4}$.

5.4 Work and energy

In order to stretch an elastic spring, a force must do work on the spring. In the case of the muscle exerciser in Figure 5.6, this force is provided by the muscles working against the tension in the spring.

When the exerciser is pulled at constant speed, at any given time the force F applied at each end is equal to the tension in the spring; consequently, it changes as the spring stretches.

Suppose that one end of the spring is stationary and the extension is x, as in Figure 5.7.

▲ **Figure 5.6**

By Hooke's law, the tension is given by:

$$T = \frac{\lambda}{l_0}x \text{ and so } F = \frac{\lambda}{l_0}x$$

▲ Figure 5.7

The work done by a *constant* force F in moving a distance d in its own direction is given by Fd. To find the work done by a variable force, the process has to be considered in small stages. Now imagine that the force extends the string by a small distance δx. The work done is given by $F\delta x = \frac{\lambda}{l_0}x\delta x$.

The work done in stretching the spring many small distances is:

$$\Sigma F\delta x = \Sigma \frac{\lambda}{l_0}x\delta x$$

When $\delta x \rightarrow 0$, the work done is:

$$\int F\mathrm{d}x = \int \frac{\lambda}{l_0}x\mathrm{d}x$$

$$= \frac{1}{2}\frac{\lambda}{l_0}x^2 + c$$

When the extension $x = 0$, the work done is zero, so $c = 0$.

The total work done in stretching the spring an extension x from its natural length l_0 is therefore given by:

$$\frac{1}{2}\frac{\lambda}{l_0}x^2$$

The result is the same for the work done in compressing a spring.

Elastic potential energy

The tensions and thrusts in perfectly elastic springs and strings are conservative forces, since any work done against them can be recovered in the form of kinetic energy. A catapult and a jack-in-a-box use this property.

The work done in stretching or compressing a string or spring can therefore be regarded as potential energy. It is known as **elastic potential energy**.

The elastic potential energy stored in a spring that is stretched or compressed by an amount x is also given by:

$$\frac{1}{2}\frac{\lambda}{l_0}x^2$$

> **Note**
>
> Notice that this is different from gravitational potential energy. Potential energy is energy that is available to be converted.

▲ Figure 5.8

Answers to exercises are available at www.hoddereducation.com/cambridgeextras

Example 5.4

An elastic rope of natural length $0.6\,\text{m}$ is extended to a length of $0.8\,\text{m}$. The modulus of elasticity of the rope is $25\,\text{N}$. Find

(i) the elastic potential energy in the rope

(ii) the further energy required to stretch it to a length of $1.65\,\text{m}$ over a car roof rack.

 Note

In this example, the elastic rope is stretched so that its extension changes from x_1 to x_2.

The work required to do this is:

$$\frac{\lambda}{2l_0}x_2^2 - \frac{\lambda}{2l_0}x_1^2 = \frac{\lambda}{2l_0}\left(x_2^2 - x_1^2\right)$$

You can see by using algebra that this expression is *not* the same as $\frac{\lambda}{2l_0}(x_2 - x_1)^2$, so it is *not* possible to use the extra extension $(x_2 - x_1)$ directly in the energy expression to calculate the extra energy stored in the elastic rope.

Solution

(i) The extension of the elastic is $(0.8 - 0.6) = 0.2\,\text{m}$.

The energy stored in the rope is $\frac{1}{2}\frac{\lambda}{l_0}x^2$

$$= \frac{25}{2 \times 0.6}\,0.2^2$$

$$= 0.833\,\text{J}$$

(ii) The extension of the elastic rope is now $1.65 - 0.6 = 1.05\,\text{m}$

The elastic energy stored in the rope is $\frac{25}{2 \times 0.6}\,1.05^2$

$$= 22.9...\,\text{J}$$

The extra energy required to stretch the rope is $22.9... - 0.833... = 22.1\,\text{J}$.

A catapult has prongs that are $16\,\text{cm}$ apart and an elastic string $20\,\text{cm}$ long.

Example 5.5

A marble of mass $70\,\text{g}$ is placed in the centre of the elastic string and pulled back so that the string is just taut. The marble is then pulled back a further $9\,\text{cm}$ and the force required to keep it in this position is $60\,\text{N}$. Find

(i) the stretched length of the string

(ii) the tension in the string and its modulus of elasticity

(iii) the elastic potential energy stored in the string and the speed of the marble when the string regains its natural length, assuming they remain in contact.

Solution

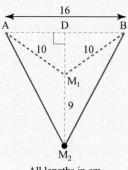

All lengths in cm

▲ **Figure 5.9**

In Figure 5.9, A and B are the ends of the elastic string and M_1 and M_2 are the two positions of the marble (before and after the string is stretched). D is the midpoint of AB.

(i)   Using Pythagoras' theorem in triangle DBM_1 gives

$$DM_1 = \sqrt{10^2 - 8^2} = 6\,\text{cm}.$$

So

$$DM_2 = 9 + 6 = 15\,\text{cm}.$$

Using Pythagoras' theorem in triangle DBM_2 gives

$$BM_2 = \sqrt{15^2 + 8^2} = 17\,\text{cm}.$$

The stretched length of the string is $2 \times 17 = 34\,\text{cm}$.

(ii)   Take the tension in the string to be T N.

Resolving parallel to M_2D:

$$2T \cos \alpha = 60$$

Now $\cos \alpha = \dfrac{DM_2}{BM_2} = \dfrac{0.15}{0.17}$

So $T = \dfrac{60 \times 0.17}{2 \times 0.15} = 34.$

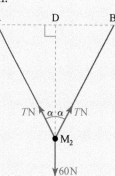

▲ **Figure 5.10**

The extension of the string is $(0.34 - 0.2) = 0.14\,\text{m}$.

By Hooke's law the modulus of elasticity λ is given by $\dfrac{\lambda}{l_0} x = T$

$$\lambda = \frac{34 \times 0.2}{0.14} = 48.5\ldots$$

The modulus of elasticity of the string is 48.6 N.

(iii)   The elastic potential energy stored in the string is:

$$\frac{1}{2}\frac{\lambda}{l_0} x^2 = \frac{1}{2} \times \frac{48.5\ldots}{0.2} \times 0.14^2 = 2.38\,\text{J}$$

By the principle of conservation of energy, this is equal to the kinetic energy given to the marble. The mass of the marble is $0.07\,\text{kg}$, so

$$\frac{1}{2} \times 0.07 v^2 = 2.38$$

$$\Rightarrow v = 8.25$$

The speed of the marble is $8.25\,\text{m s}^{-1}$.

Exercise 5C

1 An open coiled spring has natural length 0.3 m and modulus of elasticity 6 N. Find the elastic potential energy in the spring when

   (i)   it is extended by 0.1 m

   (ii)   it is compressed by 0.01 m

   (iii) its length is 0.5 m

   (iv)   its length is 0.3 m.

2 A spring has natural length 0.4 m and modulus of elasticity 20 N. Find the elastic energy stored in the spring when

   (i)   it is extended by 0.4 m

   (ii)   it is compressed by 0.1 m

   (iii) its length is 0.2 m

   (iv)   its length is 0.45 m.

3 A pinball machine fires small balls of mass 50 g by means of a spring and a light plunger. The spring and the ball move in a horizontal plane. The spring has natural length 20 cm and modulus of elasticity 120 N, and is compressed by 5 cm to fire a ball.

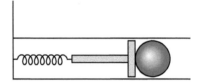

   (i)   Find the energy stored in the spring immediately before the ball is fired.

   (ii)   Find the speed of the ball when it is fired.

4 A catapult is made from elastic string with modulus of elasticity 5 N. The string is attached to two prongs that are 15 cm apart, and is just taut. A pebble of mass 40 g is placed in the centre of the string and is pulled back 4 cm and then released in a horizontal direction.

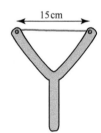

   (i)   Calculate the work done in stretching the string.

   (ii)   Calculate the speed of the pebble on leaving the catapult.

 5 A simple mathematical model of a railway buffer consists of a horizontal open coiled spring attached to a fixed point. The modulus of elasticity is 2×10^5 N and its natural length is 2 m.

The buffer is designed to stop a railway truck before the spring is compressed to half its natural length, otherwise the truck will be damaged.

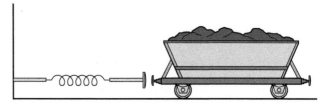

(i)   Find the elastic energy stored in the spring when it is half its natural length.

(ii) Find the maximum speed at which a truck of mass 2 tonnes can approach the buffer safely. Neglect any other reasons for loss of energy of the truck.

A truck of mass 2 tonnes approaches the buffer at $5\,\text{m s}^{-1}$.

(iii) Calculate the minimum length of the spring during the subsequent period of contact.

(iv) Find the thrust in the spring and the acceleration of the truck when the spring is at its minimum length.

(v)   What happens next?

6   Two identical springs are attached to a sphere of mass 0.5 kg that rests on a smooth horizontal surface, as shown in the diagram. The other ends of the springs are attached to fixed points A and B.

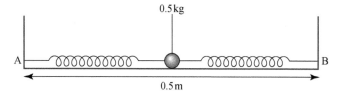

The springs each have modulus of elasticity 7.5 N and natural length 25 cm. The sphere is at rest at the midpoint when it is projected with speed $2\,\text{m s}^{-1}$ along the line of the springs towards B. Calculate the length of each spring when the sphere first comes to rest.

7   A particle P of mass 0.2 kg is attached to one end of a light elastic string of natural length 1.6 m and modulus of elasticity 18 N. The other end of the string is attached to a fixed point O which is 1.6 m above a smooth horizontal surface. P is placed on the surface vertically below O and then projected horizontally. P moves with initial speed $1.5\,\text{m s}^{-1}$ in a straight line on the surface. Show that when OP =1.8 m,

(i)   P is at instantaneous rest

(ii) P is on the point of losing contact with the surface.

Cambridge International AS & A Level Mathematics
9709 Paper 52 Q3 June 2013

8   A and B are fixed points on a smooth horizontal table. The distance AB is 2.5 m. An elastic string of natural length 0.6 m and modulus of elasticity 24 N has one end attached to the table at A, and the other end attached to a particle P of mass 0.95 kg. Another elastic string of natural length 0.9 m and modulus of elasticity 18 N has one end attached to the table at B, and the other end attached to P. The particle P is held at rest at the midpoint of AB (see diagram).

(i)   Find the tensions in the strings.

Answers to exercises are available at www.hoddereducation.com/cambridgeextras

The particle is released from rest.

(ii) Find the acceleration of P immediately after its release.

(iii) P reaches its maximum speed at the point C. Find the distance AC.

Cambridge International AS & A Level Mathematics
9709 Paper 5 Q6 June 2007

9 A particle P of mass $0.2\,\text{kg}$ is attached to one end of a light elastic string of natural length $0.8\,\text{m}$ and modulus of elasticity $64\,\text{N}$. The other end of the string is attached to a fixed point A on a smooth horizontal surface. P is placed on the surface at a point $0.8\,\text{m}$ from A. The particle P is then projected with speed $10\,\text{m s}^{-1}$ directly away from A.

(i) Calculate the distance AP when P is at instantaneous rest.

(ii) Calculate the speed of P when it is $1.0\,\text{m}$ from A.

Cambridge International AS & A Level Mathematics
9709 Paper 53 Q2 November 2014

10 A particle P of mass $1.6\,\text{kg}$ is attached to one end of each of two light elastic strings. The other ends of the strings are attached to fixed points A and B, which are $2\,\text{m}$ apart on a smooth horizontal table. The string attached to A has natural length $0.25\,\text{m}$ and modulus of elasticity $4\,\text{N}$, and the string attached to B has natural length $0.25\,\text{m}$ and modulus of elasticity $8\,\text{N}$. The particle is held at the midpoint M of AB (see diagram).

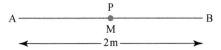

(i) Find the tensions in the strings.

(ii) Show that the total elastic potential energy in the two strings is $13.5\,\text{J}$.

P is released from rest and in the subsequent motion both strings remain taut. The displacement of P from M is denoted by $x\,\text{m}$.

(iii) Find the initial acceleration of P.

(iv) Find the non-zero value of x at which the speed of P is zero.

Cambridge International AS & A Level Mathematics
9709 Paper 5 Q6 June 2009

5.5 Vertical motion

There are two common approaches to the analysis of motion involving elastic strings and springs: using energy and using calculus. Both are covered in this section in the context of vertical motion.

Using energy

Example 5.6

A particle of mass m is attached to one end, A, of a light elastic string with natural length l_0 and modulus of elasticity λ. The other end of the string is attached to a fixed point, O. The particle is released from O.

Initially it falls freely with the string slack, but after some time it reaches a point, P, and the string becomes taut and exerts an upward force on the particle. At P, the particle has velocity u downwards. Air resistance is negligible.

(i)   Draw a diagram illustrating the situation.

(ii) Find an expression for u at the instant when the string becomes taut.

(iii) Write down the energy equation of the system at a time when the extension of the string is x m and its velocity vertically downwards is v.

(iv) Obtain an equation for the value of x when the particle is at its lowest point.

Solution

(i)

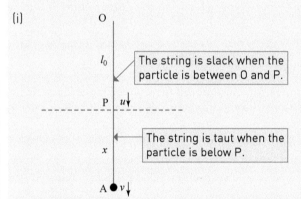

▲ **Figure 5.11**

(ii) Between O and P the particle loses gravitational potential energy mgl_0 and gains kinetic energy $\frac{1}{2}mu^2$.

Using the law of conservation of energy, $\frac{1}{2}mu^2 = mgl_0$

$$\Rightarrow u = \sqrt{2gl_0}$$

(iii) When the particle has travelled a distance x m below P

Gravitational potential energy lost $= mgx$

Elastic potential energy stored in the string $= \frac{1}{2}\left(\frac{\lambda}{l_o}\right)x^2$

Kinetic energy $= \frac{1}{2}mv^2$

Using conservation of energy:

$$\frac{1}{2}mu^2 = -mgx + \frac{1}{2}\left(\frac{\lambda}{l_o}\right)x^2 + \frac{1}{2}mv^2$$

> **Note**
>
> Notice that no units are given in this example. In such cases, you can assume that a consistent set of units is being applied to all the quantities.

> **Note**
>
> Notice that this is a quadratic equation and so will have two roots. One value is at the bottom of the particle's motion. The other would be at the top, but by then the string will have gone slack and so the model used in this example would no longer apply.

(iv) At the lowest point, the particle is stationary so $v = 0$

$$\frac{1}{2}\left(\frac{\lambda}{l_o}\right)x^2 - mgx - \frac{1}{2}mu^2 = 0$$

and since $\frac{1}{2}mu^2 = mgl_0$, this equation can be written as:

$$\frac{1}{2}\left(\frac{\lambda}{l_o}\right)x^2 - mgx - mgl_0 = 0$$

Using calculus

You can sometimes acquire more detailed information about the motion of an object by using calculus. The starting point is Newton's second law. This involves the acceleration of the body and there are three ways in which acceleration can be written.

» $\frac{dv}{dt}$ relates the velocity to the time taken

» $v\frac{dv}{ds}$ relates the velocity to the distance travelled. (It is often written as $v\frac{dv}{dx}$.)

» $\frac{d^2x}{dt^2}$ gives rise to a second-order differential equation involving derivatives of x.

Example 5.7 involves the first two of these forms. It involves exactly the same situation as the previous example, but the approaches are different.

| Example 5.7 | A particle of mass m is attached to one end, A, of a light elastic string with natural length l_0 and modulus of elasticity λ. The other end of the string is attached to a fixed point, O. The particle is released from O. |

Initially it falls freely with the string slack, but after some time it reaches a point, P, and the string becomes taut and exerts an upward force on the particle. At P, the particle has velocity u downwards. Air resistance is negligible.

(i) Write down the equation of motion for the particle when the string is taut, using $v\frac{dv}{dx}$ for the acceleration.

(ii) Solve this differential equation to find v in terms of x and interpret the solution.

(iii) Write the equation of motion for the particle when the string is taut, using $\frac{dv}{dt}$ for the acceleration, and comment on whether this is useful.

Solution

(i) Figure 5.12 shows the forces on the particle.

The equation of motion is $mg - \left(\frac{\lambda}{l_o}\right)x = ma$

or $v\frac{dv}{dx} = g - \left(\frac{\lambda}{ml_o}\right)x$

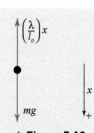

▲ Figure 5.12

(ii) Separating variables gives $\int v\,\mathrm{d}v = \int \left(g - \frac{\lambda}{l_o}x\right)\mathrm{d}x$

and so $\frac{1}{2}v^2 = gx - \frac{\lambda}{2ml_o}x^2 + c$

When $x = 0$, $v = u$, so $c = \frac{1}{2}u^2$

> These are the boundary conditions. In this case, they are also the initial conditions for the differential equation.

The solution is: $\frac{1}{2}v^2 = gx - \frac{\lambda}{2ml_o}x^2 + \frac{1}{2}u^2$.

You can interpret this in two ways:

It gives an equation for v in terms of x,

$v = \pm\sqrt{u^2 + 2gx - \frac{\lambda}{ml_o}x^2}$ or $v = \pm\sqrt{2g(l_0 + x) - \frac{\lambda}{ml_o}x^2}$ ← Replacing u^2 by $2gl_0$

With a little rearrangement, it becomes the energy equation

$\frac{1}{2}mv^2 - \frac{1}{2}mu^2 + \frac{1}{2}\frac{\lambda}{l_o}x^2 = mgx$

(iii) The equation of motion can also be written as $\frac{\mathrm{d}v}{\mathrm{d}t} = g - \frac{\lambda}{ml_o}x$.

This is not a useful form as it involves three variables, v, t and x, and so you cannot solve the differential equation.

Exercise 5D

1   A particle of mass 0.2 kg is attached to one end of a light elastic spring of modulus of elasticity 10 N and natural length 1 m. The system hangs vertically and the particle is released from rest when the spring is at its natural length. The particle comes to rest when it has fallen a distance h m.

(i) Write down an expression in terms of h for the energy stored in the spring when the particle comes to rest at its lowest point.

(ii) Write down an expression in terms of h for the gravitational potential energy lost by the particle when it comes to rest at its lowest point.

(iii) Find the value of h.

2   A particle of mass m is attached to one end of a light vertical spring of natural length l_0 and modulus of elasticity $2mg$. The particle is released from rest when the spring is at its natural length. Find, in terms of l_0, the maximum length of the spring in the subsequent motion.

3   A block of mass m is placed on a smooth plane inclined at 30° to the horizontal. The block is attached to the top of the plane by a spring of natural length l_0 and modulus of elasticity λ. The system is released from rest with the spring at its natural length. Find an expression for the maximum length of the spring in the subsequent motion.

CP 4 A particle of mass 0.1 kg is attached to one end of a spring of natural length 0.3 m and modulus of elasticity 20 N. The other end is attached to a fixed point and the system hangs vertically. The particle is released from rest when the length of the spring is 0.2 m. In the subsequent motion the extension of the spring is denoted by x m.

(i) Show that $0.05\dot{x}^2 + \frac{10}{0.3}(x^2 - 0.1^2) - (x + 0.1) = 0$.

(ii) Find the maximum value of x.

5 A small apple of mass 0.1 kg is attached to one end of an elastic string of natural length 25 cm and modulus of elasticity 5 N. Li is asleep under a tree and Huong fixes the free end of the string to the branch of the tree just above Li's head. Huong releases the apple level with the branch and it just touches Li's head in the subsequent motion. How high above her head is the branch?

6 A block of mass 0.5 kg lies on a light scale pan that is supported on a vertical spring of natural length 0.4 m and modulus of elasticity 40 N. Initially the spring is at its natural length and the block is moving downwards with a speed of 2 m s⁻¹. Gravitational potential energy is measured relative to the initial position and g should be taken to be 10 m s⁻².

(i) Find the initial mechanical energy of the system.

(ii) Show that the speed v m s⁻¹ of the block when the compression of the spring is x m is given by $v = 2\sqrt{1 + 5x - 50x^2}$.

(iii) Find the minimum length of the spring during the oscillations.

7 A scale pan of mass 0.5 kg is suspended from a fixed point by a spring of modulus of elasticity 50 N and natural length 10 cm.

(i) Calculate the length of the spring when the scale pan is in equilibrium.

(ii) A bag of sugar of mass 1 kg is gently placed on the pan and the system is released from rest. Find the maximum length of the spring in the subsequent motion.

M 8 A bungee jump is carried out by a person of mass m kg, using an elastic rope that can be taken to obey Hooke's law. The supervisor ensures that in any jump the total length of the rope will never exceed the limit of four times its original length. Prove that the tension in the rope is at most $\frac{8}{3} mg$ N.

9 A particle of mass m is attached to one end of a light elastic string with natural length l_0 and modulus of elasticity λ. The other end of the string is attached to a fixed point, O. The particle is released from O. Air resistance is negligible.

(i) Show that, at the instant when the string first becomes taut, the speed, u, of the particle is given by $u = \sqrt{2gl_0}$.

(ii) Write down the equation of motion of the particle once the string has become taut, using a to represent its acceleration and x to represent the extension of the string.

(iii) Now write the equation of motion as a differential equation, using $v\dfrac{\mathrm{d}v}{\mathrm{d}x}$ for the acceleration, where v is the velocity of the particle in the downwards direction.

(iv) Solve this differential equation and so write v in terms of x, g, m and l_0.

(v) At a certain instant, T, the velocity of the particle is given by
$$v = -\sqrt{2gl_0}.$$
Find the value of x at this time and interpret your answer.

(vi) Write down a differential equation for $\dfrac{\mathrm{d}v}{\mathrm{d}t}$ immediately after the instant T.

Solve your differential equation and state for how long your solution is valid.

10 A conical pendulum consists of a bob of mass m attached to an inextensible string of length l. The bob describes a circle of radius r with angular speed ω, and the string makes an angle θ with the vertical, as shown in the diagram.

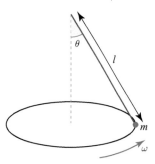

(i) Find an expression for θ in terms of ω, l and g.

The string is replaced with an elastic string of modulus of elasticity λ and natural length l_0.

(ii) Find an expression for the new value of θ in terms of ω, m, g, l_0 and λ.

11 A light elastic string has natural length 2.4 m and modulus of elasticity 21 N. A particle P of mass m kg is attached to the midpoint of the string. The ends of the string are attached to fixed points A and B which are 2.4 m apart at the same horizontal level. P is projected vertically upwards with velocity 12 m s^{-1} from the midpoint of AB. In the subsequent motion P is at instantaneous rest at a point 1.6 m above AB.

(i) Find m.

(ii) Calculate the acceleration of P when it first passes through a point 0.5 m below AB.

Cambridge International AS & A Level Mathematics
9709 Paper 51 Q4 June 2012

Answers to exercises are available at www.hoddereducation.com/cambridgeextras

12 One end of a light elastic string of natural length 0.5 m and modulus of elasticity 12 N is attached to a fixed point O. The other end of the string is attached to a particle P of mass 0.24 kg. P is projected vertically upwards with speed 3 m s⁻¹ from a position 0.8 m vertically below O.

    (i) Calculate the speed of the particle when it is moving upwards with zero acceleration.

    (ii) Show that the particle moves 0.6 m while it is moving upwards with constant acceleration.

Cambridge International AS & A Level Mathematics
9709 Paper 52 Q4 June 2011

13 A light elastic string has natural length 3 m and modulus of elasticity 45 N. A particle P of weight 6 N is attached to the midpoint of the string. The ends of the string are attached to fixed points A and B which lie in the same vertical with A above B and AB = 4 m. The particle P is released from rest at the point 1.5 m below A.

    (i) Calculate the distance P moves after its release before first coming to instantaneous rest at a point vertically above B. (You may assume that at this point the part of the string joining P to B is slack.)

    (ii) Show that the greatest speed of P occurs when it is 2.1 m below A, and calculate this greatest speed.

    (iii) Calculate the greatest magnitude of the acceleration of P.

Cambridge International AS & A Level Mathematics
9709 Paper 51 Q7 November 2012

14 One end of a light elastic string of natural length 0.4 m and modulus of elasticity 20 N is attached to a fixed point A on a smooth plane inclined at 30° to the horizontal. The other end of the string is attached to a particle P of mass 0.5 kg which rests in equilibrium on the plane.

    (i) Calculate the extension of the string.

P is projected down the plane from the equilibrium position with speed 5 m s⁻¹. The extension of the string is e m when the speed of the particle is 2 m s⁻¹ for the first time.

    (ii) Find e.

Cambridge International AS & A Level Mathematics
9709 Paper 53 Q3 June 2015

PS 15 A light elastic string of natural length 3.6 m and modulus of elasticity λ N has its ends attached to two points A and B, where AB = 3.6 m and AB is horizontal. A particle P of mass 0.5 kg is attached to the midpoint of the string. P rests in equilibrium at a distance of 0.75 m below the line AB, as shown in the diagram.

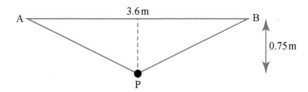

(i) Show that $\lambda = 78\,\text{N}$.

The particle is pulled downwards from its equilibrium position until the total length of the elastic string is 6 m. The particle is released from rest.

(ii) Find the speed of P when it passes the line AB.

> ## ACTIVITY 5.1
>
> ### The bungee jump
>
>
>
> ▲ **Figure 5.13**
>
> You may have noticed that the situation in Examples 5.6 and 5.7 could be used to match the bungee jump described at the start of this chapter.
>
> Typical parameters for a bungee jump:
>
> Height of jump station: 55 m
>
> Bottom safety space: 5 m
>
> Static line length: 5 m (non-elastic straps, etc.)
>
> Unstretched elastic rope length: up to 12 m
>
> Modulus of elasticity: 1000 N
>
> (i) Find the greatest mass of a jumper who can safely use the full 12 m of rope.
>
> (ii) A jumper has mass 60 kg. Find the greatest distance it is possible for her to fall.
>
> (iii) A jumper has mass 150 kg. By what length does the elastic rope need to be shortened for his jump to be safe?
>
> (iv) Find the greatest magnitude of the acceleration of a jumper of mass 80 kg.

Answers to exercises are available at www.hoddereducation.com/cambridgeextras

KEY POINTS

1 **Hooke's law**

The tension T in an elastic string or spring and its extension x are related by:

$$T = \frac{\lambda}{l_0}\, x$$

where λ is the modulus of elasticity and l_0 is the natural length of the string or spring.

2 When a spring is compressed, x is negative and the tension becomes a thrust.

3 **Elastic potential energy**

The elastic potential energy (E.P.E.) stored in a stretched spring or string, or in a compressed spring, is given

$$\text{E.P.E.} = \frac{1}{2}\,\frac{\lambda}{l_0}\,x^2$$

4 The tension or thrust in an elastic string or spring is a conservative force and so the elastic potential energy is recoverable.

5 When no frictional or other dissipative forces are involved, elastic potential energy can be used with kinetic energy and gravitational potential energy to form equations using the principle of conservation of energy.

LEARNING OUTCOMES

Now that you have finished this chapter, you should be able to

- understand the language associated with elasticity
- apply Hooke's law to strings and springs
- calculate modulus of elasticity
- find the tension in a string or spring and the thrust in a spring
- find the equilibrium position of a system involving strings or springs
- calculate the energy stored in a string or spring
- use energy to model a system involving elastic strings or springs, including determining extreme positions
- understand when Hooke's law is not applicable
- form and work with differential equations for motion under forces from elastic strings or springs.

6

Linear motion under a variable force

Is it possible to fire a projectile up to the Moon?
The Earth to the Moon, Jules Verne (1828–1905)

In his book, Jules Verne says that this is possible … 'provided it possesses an initial velocity of 12 000 yards per second. In proportion as we recede from the Earth the action of gravitation diminishes in the inverse ratio of the square of the distance; that is to say at three times a given distance the action is nine times less. Consequently the weight of a shot will decrease and will become reduced to zero at the instant that the attraction of the Moon exactly counterpoises that of the Earth; at $\frac{47}{52}$ of its journey. There the projectile will have no weight whatever; and if it passes that point it will fall into the Moon by the sole effect of lunar attraction.'

> If an *unpowered* projectile could be launched from the Earth with a high enough speed in the right direction, it would reach the Moon.
>
> ❯ What forces act on the projectile during its journey?
>
> ❯ How near to the Moon will it get if its initial speed is not *quite* enough?

In Jules Verne's story, three men and two dogs were sent to the Moon inside a projectile fired from an enormous gun. Although this is completely impracticable, the basic mathematical ideas in the passage quoted on the previous page are correct. As a projectile moves further from the Earth and nearer to the Moon, the gravitational attraction of the Earth decreases and that of the Moon increases. In many of the dynamics problems you have met so far it has been assumed that forces are *constant*, whereas on Jules Verne's space missile the total force *varies* continuously as the motion proceeds.

You may have already met problems involving variable force. When an object is suspended on a spring and bounces up and down, the varying tension in the spring leads to **simple harmonic motion**. You will also be aware that air resistance depends on velocity.

Gravitation, spring tension and air resistance all give rise to variable force problems; the subject of this chapter.

6.1 Newton's second law as a differential equation

Calculus techniques are used extensively in mechanics and you will already have used differentiation and integration in earlier work. In this chapter you will see how essential calculus methods are in the solution of a variety of problems.

To solve variable force problems, you can use Newton's second law to give an equation for the *instantaneous* value of the acceleration. When the mass of a body is constant, this can be written in the form of a **differential equation**.

$$F = m\frac{\mathrm{d}v}{\mathrm{d}t}$$

It can also be written as

$$F = mv\frac{\mathrm{d}v}{\mathrm{d}s}$$

> This formula is also often written as $F = m\dfrac{\mathrm{d}^2 s}{\mathrm{d}t^2}$.

This follows from the chain rule for differentiation.

$$\frac{\mathrm{d}v}{\mathrm{d}t} = \frac{\mathrm{d}v}{\mathrm{d}s} \times \frac{\mathrm{d}s}{\mathrm{d}t}$$

$$= v\frac{\mathrm{d}v}{\mathrm{d}s}$$

 Note

Here and throughout this chapter the mass, m, is assumed to be constant. Jules Verne's spacecraft was a projectile fired from a gun. It was not a *rocket*, with mass that varies due to ejection of fuel.

Deriving the constant acceleration formulae

To see the difference in use between the $\dfrac{\mathrm{d}v}{\mathrm{d}t}$ and $v\dfrac{\mathrm{d}v}{\mathrm{d}s}$ forms of acceleration, it is worth looking at the case where the force, and therefore the acceleration, $\dfrac{F}{m}$ is *constant* (say a). Starting from the $\dfrac{\mathrm{d}v}{\mathrm{d}t}$ form,

$$\frac{\mathrm{d}v}{\mathrm{d}t} = a$$

Integrating gives

$$v = u + at$$

where u is the constant of integration.

$v = u$ when $t = 0$.

Since $v = \dfrac{\mathrm{d}s}{\mathrm{d}t}$, integrating again gives

$$s = ut + \tfrac{1}{2}at^2 + s_0$$

Assuming the displacement is s_0 when $t = 0$.

These are the familiar formulae for motion under constant acceleration.

Starting from the $v\dfrac{\mathrm{d}v}{\mathrm{d}s}$ form,

$$v\frac{\mathrm{d}v}{\mathrm{d}s} = a$$

Separating the variables and integrating gives

$$\int v\,\mathrm{d}v = \int a\,\mathrm{d}s$$

$$\Rightarrow \qquad \tfrac{1}{2}v^2 = as + k$$

where k is the constant of integration.

Assuming $v = u$ when $s = 0$, $k = \tfrac{1}{2}u^2$.

So the formula becomes $v^2 = u^2 + 2as$

This is another of the standard constant acceleration formulae. Notice that *time is not involved* when you start from the $v\dfrac{\mathrm{d}v}{\mathrm{d}s}$ form of acceleration.

Solving $F = ma$ for variable force

When the force is continuously *variable*, you write Newton's second law in the form of a differential equation and then solve it, using one of the forms of acceleration, $v\dfrac{\mathrm{d}v}{\mathrm{d}s}$ or $\dfrac{\mathrm{d}v}{\mathrm{d}t}$. The choice depends on the particular problem. Some guidelines are given below and you should check these with the examples that follow.

Normally, the resulting differential equation can be solved by separating the variables.

When the force is a function of time

When the force is a function, $F(t)$, of time you use $a = \dfrac{\mathrm{d}v}{\mathrm{d}t}$.

$$F(t) = m\frac{\mathrm{d}v}{\mathrm{d}t}$$

Answers to exercises are available at www.hoddereducation.com/cambridgeextras

Separating the variables and integrating gives

$$m \int dv = \int F(t)\, dt$$

Assuming you can solve the integral on the right-hand side, you then have v in terms of t.

Writing v as $\dfrac{ds}{dt}$, you can find the displacement as a function of time by integrating again.

When the force is a function of displacement

When the force is a function, $F(s)$, of displacement, you normally start from

$$F(s) = mv \dfrac{dv}{ds}$$

then $\qquad \int F(s)\, ds = m \int v\, dv.$

When the force is a function of velocity

When the force is given as a function, $F(v)$, of velocity, you have a choice. You can use

$$F(v) = m \dfrac{dv}{dt}$$

or $\qquad\qquad F(v) = mv \dfrac{dv}{ds}$

then $\qquad\qquad s = m \int v \dfrac{dv}{F(v)}$

You can separate the variables in both forms; use the first if you are interested in behaviour over time and the second when you wish to involve displacement.

6.2 Variable force examples

Examples 6.1–6.3 show the approaches used when the force is given respectively as a function of time, displacement and velocity.

When you are solving these problems, it is important to be clear about which direction is positive *before* writing down an equation of motion.

Example 6.1

A crate of mass m is freely suspended at rest from a crane. When the operator begins to lift the crate further, the tension in the suspending cable increases uniformly from mg newtons to $1.2\, mg$ newtons over a period of 2 seconds.

(i)   What is the tension in the cable t seconds after the lifting has begun ($t \leqslant 2$)?

(ii)   What is the velocity after 2 seconds?

(iii)   How far has the crate risen after 2 seconds?

Assume the situation may be modelled with air resistance and cable stretching ignored.

Solution

When the crate is at rest it is in equilibrium and so the tension, T, in the cable equals the weight mg of the crate. After time $t = 0$, the tension increases, so there is a net upward force and the crate rises, see Figure 6.1.

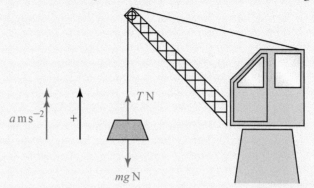

▲ **Figure 6.1**

(i) The tension increases uniformly by $0.2mg$ newtons in 2 seconds, which means that it increases by $0.1mg$ newtons per second (see Figure 6.2).

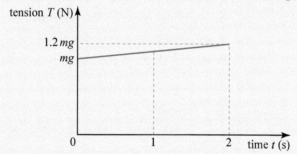

▲ **Figure 6.2**

After t seconds, the tension is $T = mg + 0.1mgt$.

(ii) As the force is a function of time, use $a = \dfrac{\mathrm{d}v}{\mathrm{d}t}$. Then at any moment in the 2-second period, $F = ma$ gives

$$(mg + 0.1mgt) - mg = m\frac{\mathrm{d}v}{\mathrm{d}t} \qquad \boxed{\text{Upwards is positive.}}$$

$$\Rightarrow \qquad\qquad\qquad \frac{\mathrm{d}v}{\mathrm{d}t} = 0.1gt$$

Integrating gives

$$v = 0.05gt^2 + k \qquad \boxed{\begin{array}{l} k \text{ is the constant} \\ \text{of integration.} \end{array}}$$

When $t = 0$, the crate has not quite begun to move, so $v = 0$. This gives $k = 0$ and $v = 0.05gt^2$.

When t is 2,

$$v = 0.05 \times 10 \times 4$$
$$= 2.$$

The velocity after 2 seconds is $2\,\mathrm{m\,s^{-1}}$.

Answers to exercises are available at www.hoddereducation.com/cambridgeextras

(iii) To find the displacement s, write v as $\dfrac{ds}{dt}$ and integrate again.

$$\frac{ds}{dt} = 0.05gt^2$$

$$s = \int 0.05gt^2\,dt$$

$$s = 0.05g \times \tfrac{1}{3}t^3 + c \quad \longleftarrow \quad \boxed{\text{When } t = 0,\, s = 0 \Rightarrow c = 0.}$$

When $t = 2$ and $g = 10$, $s = \tfrac{4}{3}$.

The crate moves $\tfrac{4}{3}$ m in 2 seconds.

> The displacement cannot be obtained by the formula $s = \tfrac{1}{2}(u + v)t$, which would give the answer $2\,\text{m}$. Why not?

Example 6.2

A prototype of Jules Verne's projectile, mass m, is launched vertically upwards from the Earth's surface, but only just reaches a height of one tenth of the Earth's radius before falling back. When the height, s, above the surface is small compared with the radius, R, of the Earth, the magnitude of the Earth's gravitational force on the projectile may be modelled as $mg\left(1 - \dfrac{2s}{R}\right)$, where g is gravitational acceleration at the Earth's surface.

Assuming all other forces can be neglected

(i) write down a differential equation of motion involving s and velocity, v

(ii) integrate this equation and hence obtain an expression for the loss of kinetic energy of the projectile between its launch and it rising to a height s

(iii) show that the launch velocity is $0.3\sqrt{2gR}$.

Solution

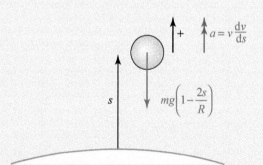

▲ **Figure 6.3**

(i) Taking the upward direction as positive, the force on the projectile is $-mg\left(1 - \dfrac{2s}{R}\right)$. The force is a function of s, so start from the equation of motion in the form

$$mv\frac{dv}{ds} = -mg\left(1 - \frac{2s}{R}\right).$$

(ii) Separating the variables and integrating gives

$$\int mv \, dv = -\int mg \left(1 - \frac{2s}{R}\right) ds$$

$$\Rightarrow \qquad \frac{1}{2}mv^2 = -mgs + \frac{mgs^2}{R} + k.$$

> You would normally divide the equation by m, but it is useful to leave it in here in order to get kinetic energy directly from $\int mv \, dv$.

Writing v_0 for the launch velocity, $v = v_0$ when $s = 0$, so $k = \frac{1}{2}mv_0^2$ and rearranging gives

$$\frac{1}{2}mv_0^2 - \frac{1}{2}mv^2 = mgs - \frac{mgs^2}{R}. \qquad \text{①}$$

The left-hand side is the loss of kinetic energy, so

$$\text{loss of K.E.} = mgs - \frac{mgs^2}{R}.$$

(iii) Dividing equation ① by m and multiplying by 2 gives

$$v_0^2 - v^2 = 2gs - \frac{2gs^2}{R}.$$

If the projectile just reaches a height $s = \frac{R}{10}$, then the velocity v is zero at that point.

Substituting $s = \frac{R}{10}$ and $v = 0$ gives

$$v_0^2 = 2g\left(\frac{R}{10}\right) - \frac{2gR^2}{100R}$$

$$= \frac{18gR}{100}$$

$$\Rightarrow \qquad v_0 = \frac{3}{10}\sqrt{2gR}$$

So the launch velocity is $0.3\sqrt{2gR}$.

Example 6.3

A body of mass $2\,\text{kg}$, initially at rest on a smooth horizontal plane, is subjected to a horizontal force of magnitude $\frac{1}{2v+1}\,\text{N}$, where v is the velocity of the body ($v > 0$).

(i) Find the time at which the velocity is $1\,\text{m s}^{-1}$.

(ii) Find the displacement when the velocity is $1\,\text{m s}^{-1}$.

Solution

(i) Using $F = ma = m\dfrac{dv}{dt}$

> Write acceleration in $\dfrac{dv}{dt}$ form, since time is required.

$$\frac{1}{2v+1} = 2\frac{dv}{dt}.$$

Separating the variables gives

$$\int dt = \int 2(2v + 1) \, dv$$

$$\Rightarrow \qquad t = 2v^2 + 2v + k.$$

> When $t = 0$, $v = 0$, so $k = 0$.

Therefore

$$t = 2v^2 + 2v$$

When $v = 1$, $t = 4$. That is, when the velocity is $1\,\mathrm{m\,s^{-1}}$, the time is 4 seconds.

(ii) Using $F = ma = mv\dfrac{\mathrm{d}v}{\mathrm{d}s}$

$$\frac{1}{2v+1} = 2v\frac{\mathrm{d}v}{\mathrm{d}s}$$

> Write acceleration in $v\dfrac{\mathrm{d}v}{\mathrm{d}s}$ form, since displacement is required.

Separating the variables gives

$$\int \mathrm{d}s = \int 2v(2v + 1)\,\mathrm{d}v$$

$$\Rightarrow \qquad s = \tfrac{4}{3}v^3 + v^2 + k. \quad \longleftarrow \boxed{\text{When } s = 0,\, v = 0,\text{ so } k = 0.}$$

Therefore

$$s = \tfrac{4}{3}v^3 + v^2$$

When $v = 1$, $s = \tfrac{7}{3}$.

When the velocity is $1\,\mathrm{m\,s^{-1}}$, the displacement is $2\tfrac{1}{3}\,\mathrm{m}$.

Exercise 6A

1 Each of the parts (i) to (viii) of this question assumes a body of mass $1\,\mathrm{kg}$ under the influence of a single force F N acting in a constant direction but with a variable magnitude, given as a function of velocity, $v\,\mathrm{m\,s^{-1}}$, displacement, $s\,\mathrm{m}$, or time, t seconds.

In each case, express $F = ma$ as a differential equation, using either $a = \dfrac{\mathrm{d}v}{\mathrm{d}t}$ or $a = v\dfrac{\mathrm{d}v}{\mathrm{d}s}$ as appropriate. Then separate the variables and integrate, giving the result in the required form and leaving an arbitrary constant in the answer.

(i) $F = 2v$ express s in terms of v

(ii) $F = 2v$ express v in terms of t

(iii) $F = 2\sin 3t$ express v in terms of t

(iv) $F = -v^2$ express v in terms of t

(v) $F = -v^2$ express v in terms of s

(vi) $F = -4s + 2$ express v in terms of s

(vii) $F = -2v - 3v^2$ express s in terms of v

(viii) $F = 1 + v^2$ express s in terms of v

2 Each of the parts (i) to (viii) of this question assumes a body of mass $1\,\mathrm{kg}$ under the influence of a single force F N acting in a constant direction but with a variable magnitude, given as a function of velocity, $v\,\mathrm{m\,s^{-1}}$, displacement, $s\,\mathrm{m}$, or time, t seconds. The body is initially at rest at a point O.

In each case, write down the equation of motion and solve it to supply the required information.

(i) $F = 2t^2$ find v when $t = 2$

(ii) $F = -\dfrac{1}{(s+1)^2}$ find v when $s = -\frac{1}{9}$

(iii) $F = \dfrac{1}{s+3}$ find v when $s = 3$

(iv) $F = \dfrac{1}{v+1}$ find t when $v = 3$

(v) $F = 1 + v^2$ find t when $v = 1$

(vi) $F = 5 - 3v$ find t when $v = 1$

(vii) $F = 1 - v^2$ find t when $v = 0.5$ **Tip:** Use partial fractions.

(viii) $F = 1 - v^2$ find s when $v = 0.5$

3   A horse pulls a $500\,\text{kg}$ cart from rest until the speed, v, is about $5\,\text{m s}^{-1}$. Over this range of speeds, the magnitude of the force exerted by the horse can be modelled by $500(v + 2)^{-1}\,\text{N}$. Assume that resistance can be neglected.

(i)   Write down an expression for $v\dfrac{\mathrm{d}v}{\mathrm{d}s}$ in terms of v.

(ii)   Show by integration that when the velocity is $3\,\text{m s}^{-1}$, the cart has travelled $18\,\text{m}$.

(iii)   Write down an expression for $\dfrac{\mathrm{d}v}{\mathrm{d}t}$ and integrate to show that the velocity is $3\,\text{m s}^{-1}$ after 10.5 seconds.

(iv)   Show that $v = -2 + \sqrt{4 + 2t}$.

(v)   Integrate again to derive an expression for s in terms of t, and verify that, after 10.5 seconds, the cart has travelled $18\,\text{m}$.

4   A particle P of mass $0.5\,\text{kg}$ is projected vertically upwards from a point on a horizontal surface. A resisting force of magnitude $0.02v^2\,\text{N}$ acts on P where $v\,\text{m s}^{-1}$ is the upward velocity of P when it is a height of $x\,\text{m}$ above the surface. The initial speed of P is $8\,\text{m s}^{-1}$.

(i)   Show that, while P is moving upwards, $v\dfrac{\mathrm{d}v}{\mathrm{d}x} = -10 - 0.04v^2$.

(ii)   Find the greatest height of P above the surface.

(iii)   Find the speed of P immediately before it strikes the surface after descending.

Cambridge International AS & A Level Mathematics
9709 Paper 53 Q5 November 2015

5   An object of mass $0.4\,\text{kg}$ is projected vertically upwards from the ground, with an initial speed of $16\,\text{m s}^{-1}$. A resisting force of magnitude $0.1v\,\text{N}$ acts on the object during its ascent, where $v\,\text{m s}^{-1}$ is the speed of the object at time ts after it starts to move.

(i)   Show that $\dfrac{\mathrm{d}v}{\mathrm{d}t} = -0.25(v + 40)$.

(ii)   Find the value of t at the instant that the object reaches its maximum height.

Cambridge International AS & A Level Mathematics
9709 Paper 5 Q4 June 2006

Answers to exercises are available at www.hoddereducation.com/cambridgeextras

6   A particle P of mass 0.5 kg moves on a horizontal surface along the straight line OA, in the direction from O to A. The coefficient of friction between P and the surface is 0.08. Air resistance of magnitude $0.2v$ N opposes the motion, where $v\,\mathrm{m\,s^{-1}}$ is the speed of P at time t s. The particle passes through O with speed $4\,\mathrm{m\,s^{-1}}$ when $t = 0$.

  (i)   Show that $2.5\dfrac{\mathrm{d}v}{\mathrm{d}t} = -(v + 2)$ and hence find the value of t when $v = 0$.

  (ii) Show that $\dfrac{\mathrm{d}x}{\mathrm{d}t} = 6\mathrm{e}^{-0.4t} - 2$, where x m is the displacement of P from O at time t s, and hence find the distance OP when $v = 0$.

Cambridge International AS & A Level Mathematics
9709 Paper 5 Q7 June 2008

7   A particle P starts from a fixed point O and moves in a straight line. When the displacement of P from O is x m, its velocity is $v\,\mathrm{m\,s^{-1}}$ and its acceleration is $\dfrac{1}{x+2}\,\mathrm{m\,s^{-2}}$.

  (i)   Given that $v = 2$ when $x = 0$, use integration to show that $v^2 = 2\ln\!\left(\tfrac{1}{2}x + 1\right) + 4$.

  (ii) Find the value of v when the acceleration of P is $\tfrac{1}{4}\,\mathrm{m\,s^{-2}}$.

Cambridge International AS & A Level Mathematics
9709 Paper 5 Q3 June 2009

8   A particle of mass 0.4 kg is released from rest at the top of a smooth plane inclined at 30° to the horizontal. The motion of P down the slope is opposed by a force of magnitude $0.6x$ N, where x m is the distance P has travelled down the slope. P comes to rest before reaching the foot of the slope. Calculate

  (i)   the greatest speed of P during its motion

  (ii) the distance travelled by P during its motion.

Cambridge International AS & A Level Mathematics
9709 Paper 51 Q5 June 2012

9   A particle P of mass 0.5 kg moves in a straight line on a smooth horizontal surface. The velocity of P is $v\,\mathrm{m\,s^{-1}}$ when the displacement of P from O is x m. A single horizontal force of magnitude $0.16\mathrm{e}^x$ N acts on P in the direction OP. The velocity of P when it is at O is $0.8\,\mathrm{m\,s^{-1}}$.

  (i)   Show that $v = 0.8\mathrm{e}^{\frac{1}{2}x}$.

  (ii) Find the time taken by P to travel 1.4 m from O.

Cambridge International AS & A Level Mathematics
9709 Paper 52 Q7 June 2013

10 A cyclist and her bicycle have a total mass of 60 kg. The cyclist rides in a horizontal straight line, and exerts a constant force in the direction of motion of 150 N. The motion is opposed by a resistance of magnitude $12v$ N, where $v\,\mathrm{m\,s^{-1}}$ is the cyclist's speed at time t s after passing through a fixed point A.

  (i)   Show that $5\dfrac{\mathrm{d}v}{\mathrm{d}t} = 12.5 - v$.

(ii) Given that the cyclist passes through A with speed $11.5\,\text{m}\,\text{s}^{-1}$, solve this differential equation to show that $v = 12.5 - e^{-0.2t}$.

(iii) Express the displacement of the cyclist from A in terms of t.

Cambridge International AS & A Level Mathematics
9709 Paper 53 Q6 June 2015

11 A particle P of mass $0.3\,\text{kg}$ is projected vertically upwards from the ground with an initial speed of $20\,\text{m}\,\text{s}^{-1}$. When P is at height $x\,\text{m}$ above the ground, its upward speed is $v\,\text{m}\,\text{s}^{-1}$. It is given that

$$3v - 90\ln(v + 30) + x = A$$

where A is a constant.

(i) Differentiate this equation with respect to x and hence show that the acceleration of the particle is $-\frac{1}{3}(v + 30)\,\text{m}\,\text{s}^{-2}$.

(ii) Find, in terms of v, the resisting force acting on the particle.

(iii) Find the time taken for P to reach its maximum height.

Cambridge International AS & A Level Mathematics
9709 Paper 52 Q7 November 2009

12 A particle P of mass $0.25\,\text{kg}$ moves in a straight line on a smooth horizontal surface. P starts at the point O with speed $10\,\text{m}\,\text{s}^{-1}$ and moves towards a fixed point A on the line. At time $t\,\text{s}$ the displacement of P from O is $x\,\text{m}$ and the velocity of P is $v\,\text{m}\,\text{s}^{-1}$. A resistive force of magnitude $(5 - x)\,\text{N}$ acts on P in the direction towards O.

(i) Form a differential equation in v and x. By solving this differential equation, show that $v = 10 - 2x$.

(ii) Find x in terms of t, and hence show that the particle is always less than $5\,\text{m}$ from O.

Cambridge International AS & A Level Mathematics
9709 Paper 51 Q7 June 2010

PS 13 A rocket of mass $1000\,\text{kg}$ is launched from rest at ground level and travels vertically upwards. The mass of the rocket is constant and the only forces acting on it are its weight, a driving force of $20\,000\,\text{N}$ and a resistance force $5v\,\text{N}$.

(i) Show that $\dfrac{\text{d}v}{\text{d}t} = 10 - 0.005v$.

(ii) Find v in terms of t.

(iii) Find the distance travelled by the rocket in the first $5\,\text{s}$ of its motion.

PS 14 A particle is projected with speed U at time $t = 0$ and moves in a straight line. At time t, its velocity is v and the distance travelled is x. The acceleration of the particle is $-k\sqrt{v}$ where k is a constant.

(i) Show that the particle will come to rest when $t = \dfrac{2\sqrt{U}}{k}$.

(ii) Find, in terms of k and U, the distance travelled while the particle comes to rest.

15 A particle of mass 0.4 kg is projected vertically upwards with a speed of $20 \, \text{m s}^{-1}$. The particle experiences a resistance of $0.5v \, \text{N}$, where $v \, \text{m s}^{-1}$ is the velocity of the particle.

(i) Find the time taken for the particle to come to instantaneous rest.

(ii) Find the greatest height attained by the particle above its point of projection.

Having reached its highest point, the particle then drops down against a resistance of $0.5v \, \text{N}$.

(iii) Show that $\dfrac{\text{d}v}{\text{d}t} = 1.25(8 - v)$ and use it to find an expression for v as a function of t, where t is the time in seconds that has elapsed since the particle reached its *highest point*.

(iv) Find an expression for the distance travelled by the particle *during its descent* as a function of t. Hence show that the time taken for the particle to drop down to ground level is greater than the time taken to reach its highest point.

KEY POINTS

1 When a particle is moving along a line under a variable force F, Newton's second law gives a differential equation. It is generally solved by writing acceleration as

$\dfrac{\text{d}v}{\text{d}t}$ when F is given as a function of time, t

$v\dfrac{\text{d}v}{\text{d}s}$ when F is given as a function of displacement, s

$\dfrac{\text{d}v}{\text{d}t}$ or $v\dfrac{\text{d}v}{\text{d}s}$ when F is given as a function of velocity, v.

LEARNING OUTCOMES

Now that you have finished this chapter, you should be able to

■ solve problems that can be modelled as the linear motion of a particle under the action of a variable force

■ set up and solve a differential equation corresponding to Newton's second law

■ use the form for acceleration

■ solve differential equations in which the variables are separable.

7 Momentum

?

> The karate expert in the photograph has just broken a pile of wooden planks with a single blow from his hand. Forces in excess of 3000 N have been measured during karate chops. How is this possible?

e

Note

This chapter is about momentum and collisions. The opening section uses the concept of impulse to link the new ideas involved to familiar work on Newton's second law. You can, however, answer typical questions on impact without using impulse. The aim of Section 7.1 is to improve your understanding of the whole chapter.

7.1 Impulse

Although the karate expert in the photograph produces a very large force, it acts for only a short time. This is often the case in situations where impacts occur, as in the following example involving a tennis player.

Example 7.1

A tennis player hits the ball as it is travelling towards her at $10\,\mathrm{m\,s^{-1}}$ horizontally. Immediately after she hits it, the ball is travelling away from her at $20\,\mathrm{m\,s^{-1}}$ horizontally. The mass of the ball is $0.06\,\mathrm{kg}$. What force does the tennis player apply to the ball?

Solution

You cannot tell unless you know how long the impact lasts, and that will vary from one shot to another.

Although you cannot calculate the force unless you know the time for which it acts, you can work out the product force × time. This is called the **impulse**. An impulse is usually denoted by **J** and its magnitude by *J*.

> ❯ For the tennis player in Example 7.1, show that the average force she applies to the ball in the cases where the impact lasts 0.1 s and 0.015 s are 18 N and 120 N, respectively. What does 'average' mean in this context?

When a constant force acts for a time *t* the impulse of the force is defined as

$$\text{impulse} = \text{force} \times \text{time}.$$

The impulse is a vector acting in the direction of the force. Impulse is often used when force and time cannot be known separately but their combined effect is known, as in the case of the tennis ball. The S.I. unit for impulse is the Newton second (N s).

Impulse and momentum

When the motion is in one dimension and the velocity of an object of mass *m* is changed from *u* to *v* by a constant force *F*, you can use Newton's second law and the equations for motion with constant acceleration.

$$F = ma$$

and

$$v = u + at$$

$$\Rightarrow \quad mv = mu + mat$$

Substituting *F* for *ma* gives

$$mv = mu + Ft$$

$$\Rightarrow \quad Ft = mv - mu \qquad \text{①}$$

The quantity 'mass × velocity' is defined as the **momentum** of the moving object.

The equation ① can be written as

$$\text{impulse of force} = \text{final momentum} - \text{initial momentum.} \qquad \text{②}$$

So impulse = change in momentum.

final momentum

initial momentum

The −10 takes account of the change in direction.

This equation also holds for any large force acting for a short time, even when it cannot be assumed to be constant. The force on the tennis ball will increase as it embeds itself into the strings and then decrease as it is catapulted away, but you can calculate the impulse of the tennis racket on the ball as

$$0.06 \times 20 - 0.06 \times (-10) = 1.8 \, \text{N s.}$$

impulse

Equation ② is also true for a variable force. It is also true, but less often used, when a longer time is involved.

| Example 7.2 | A ball of mass 50 g hits the ground with a speed of $4\,\mathrm{m\,s^{-1}}$ and rebounds with an initial speed of $3\,\mathrm{m\,s^{-1}}$. The situation is modelled by assuming that the ball is in contact with the ground for 0.01 s and that during this time the reaction force on it is constant. |

(i)   Find the average force exerted on the ball by the ground.

(ii) Find the loss in kinetic energy during the impact.

(iii) Which of the answers to parts (i) and (ii) would be affected by a change in the modelling assumption that the ball is only in contact with the ground for 0.01 s?

Solution

(i) The impulse is given by:

$$J = mv - mu$$

$$= 0.05 \times 3 - 0.05 \times (-4)$$

$$= 0.35$$

The impulse is also given by

$$J = Ft$$

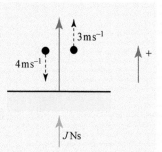
▲ Figure 7.1

where F is the average force, i.e. the constant force which, acting for the same time interval, would have the same effect as the variable force which actually acted.

$$\therefore \qquad 0.35 = F \times 0.01$$

$$F = 35$$

So the ground exerts an average upward force of 35 N.

(ii) Initial K.E. $= \frac{1}{2} \times 0.05 \times 4^2$

$$= 0.400\,\mathrm{J}$$

Final K.E. $= \frac{1}{2} \times 0.05 \times 3^2$

$$= 0.225\,\mathrm{J}$$

Loss in K.E. $= 0.175\,\mathrm{J}$

(This is converted into heat and sound.)

(iii) A change in the model will affect the answer to part (i), but not part (ii).

 Note

This example demonstrates the important point that mechanical energy is not conserved during an impact.

Although the force of gravity acts during the impact, its impulse is negligible over such a short time.

Answers to exercises are available at www.hoddereducation.com/cambridgeextras

> The magnitude of the momentum of an object is often thought of as its resistance to being stopped. Compare the momentum and kinetic energy of a cricket ball of mass 0.15 kg bowled at a high speed of $40\,\mathrm{m\,s^{-1}}$ and a 20 tonne railway truck moving at the very low speed of 1 cm per second.

> Which would you rather be hit by, an object with high momentum and low speed, or one with high speed and low momentum?

Example 7.3

A car of mass 800 kg is pushed with a constant force of magnitude 200 N for 10 s. The car starts from rest. Resistance to motion may be ignored.

(i)   Find its speed at the end of the ten-second interval by using

(a) the impulse on the car   (b) Newton's second law.

(ii) Comment on your answers to part (i).

Solution

(i)   (a) The force of 200 N acts for 10 s, so the impulse on the car is

| The impulse is in the direction of the force. |

$$J = 200 \times 10 = 2000\,\mathrm{N\,s}$$

Hence the change in momentum (in N s) is

$$mv = 2000$$

| Since the force is assumed to be constant, so is the acceleration and hence you can use the constant acceleration formulae. |

$$\therefore \qquad v = \frac{2000}{800} = 2.5$$

The speed at the end of the time interval is $2.5\,\mathrm{m\,s^{-1}}$.

(b) Newton's second law

$$F = ma$$

$$200 = 800a$$

$$a = 0.25\ \mathrm{m\,s^{-2}}$$

$$v = u + at$$

$$v = 0 + 0.25 \times 10 = 2.5\ \mathrm{m\,s^{-1}}$$

(ii) Both methods give the same answer but the method based on Newton's second law and the constant acceleration formulae only works because the force is constant.

Consider a variable force $F(t)$ acting on an object in the interval of time $0 \leqslant t \leqslant T$, which changes its velocity from U to V.

At any instant, Newton's second law gives

$$F = ma = m\,\frac{\mathrm{d}v}{\mathrm{d}t}$$

and so the overall effect is given by

$$\int_0^T F\,dt = m\int_{v=U}^{v=V}\frac{dv}{dt}\,dt = m\int_U^V dv$$
$$= mV - mU$$

This is the impulse–momentum equation.

Exercise 7A	

1 Find the momentum of these objects, assuming each of them to be travelling in a straight line.

(i) An ice skater of mass 50 kg travelling with speed $10\,\mathrm{ms^{-1}}$.

(ii) An elephant of mass 5 tonnes moving at $4\,\mathrm{ms^{-1}}$.

(iii) A train of mass 7000 tonnes travelling at $40\,\mathrm{ms^{-1}}$.

(iv) A bacterium of mass $2 \times 10^{-16}\,\mathrm{g}$ moving with speed $1\,\mathrm{mms^{-1}}$.

2 Calculate the impulse required in each of these situations:

(i) to stop a car of mass 1.3 tonnes travelling at $14\,\mathrm{ms^{-1}}$

(ii) to putt a golf ball of mass 1.5 g with speed $1.5\,\mathrm{ms^{-1}}$

(iii) to stop a cricket ball of mass 0.15 kg travelling at $20\,\mathrm{ms^{-1}}$

(iv) to fire a bullet of mass 25 g with speed $400\,\mathrm{ms^{-1}}$.

3 A stone of mass 1.5 kg is dropped from rest. After a time interval t s, it has fallen a distance s m and has velocity $v\,\mathrm{ms^{-1}}$.

Take g to be $10\,\mathrm{ms^{-2}}$ and neglect air resistance.

(i) Write down the force F (in N) acting on the stone.

(ii) Find s when $t = 2\,s$.

(iii) Find v when $t = 2\,s$.

(iv) Write down the value, units and meaning of Fs when $t = 2\,s$ and explain why this has the same value as $\frac{1}{2} \times 1.5v^2$ when $t = 2\,s$.

(v) Write down the value, units and meaning of Ft when $t = 2\,s$ and explain why this has the same value as $1.5v$ when $t = 2\,s$.

4 A ball of mass 200 g is moving in a straight line with a speed of $5\,\mathrm{ms^{-1}}$ when a force of 20 N is applied to it for 0.1 s in the direction of motion. Find the final speed of the ball

(i) (a) using the impulse–momentum equation

(b) using Newton's second law and the constant acceleration formulae.

(ii) Compare the methods.

5 A girl throws a ball of mass 0.06 kg vertically upwards with initial speed $20\,\mathrm{ms^{-1}}$. Take g to be $10\,\mathrm{ms^{-2}}$ and neglect air resistance.

(i) What is the initial momentum of the ball?

(ii) How long does it take for the ball to reach the top of its flight?

(iii) What is the momentum of the ball when it is at the top of its flight?

(iv) What impulse acted on the ball over the period between its being thrown and its reaching maximum height?

6 A netball of mass 425 g is moving horizontally with speed $5\,\text{m s}^{-1}$ when it is caught.

(i) Find the impulse needed to stop the ball.

(ii) Find the average force needed to stop the ball if it takes

    (a) 0.1 s   (b) 0.05 s.

(iii) Why does the action of 'cushioning' the ball with your hands make it easier to catch?

7 A car of mass 0.9 tonnes is travelling at $13.2\,\text{m s}^{-1}$ when it crashes head-on into a wall. The car is brought to rest in a time of 0.12 s. Find

(i)   (a) the impulse acting on the car

    (b) the average force acting on the car

    (c) the average deceleration of the car.

(ii) Explain why many cars are designed with crumple zones rather than with completely rigid construction.

8 Boris is sleeping on a bunk bed at a height of 1.5 m when he rolls over and falls out. His mass is 20 kg. Find

(i) the speed with which he hits the floor

(ii) the impulse that the floor has exerted on him when he has come to rest

(iii) the impulse he has exerted on the floor.

It takes Boris 0.2 s to come to rest.

(iv) Find the average force acting on him during this time.

9 A railway truck of mass 10 tonnes is travelling at $3\,\text{m s}^{-1}$ along a siding when it hits some buffers. After the impact it is travelling at $1.5\,\text{m s}^{-1}$ in the opposite direction.

(i) Find the initial momentum of the truck.

(ii) Find the momentum of the truck after it has left the buffers.

(iii) Find the impulse that has acted on the truck.

During the impact the force $F\,\text{N}$ that the buffers exert on the truck varies as shown in this graph.

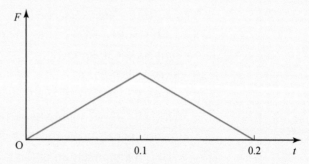

(iv) State what information is given by the area under the graph.

(v) What is the greatest value of the force F?

10 A van of mass $2500\,\text{kg}$ starts from rest. In the first 4 seconds after starting, the driving force on its engine follows the relationship $F(t) = 2400t - 300t^2$. Find

(i)   the total impulse on the van over the 4 seconds

(ii) the speed of the van after 4 seconds, ignoring the effects of air resistance or friction.

7.2 Conservation of momentum

Collisions

The law of conservation of momentum plays an important role in the consideration of collision problems.

It states that when there are no external influences on a system the total momentum of the system remains constant.

Suppose, for example, two objects A and B collide while moving along a straight line. A is initially moving with speed u_A and is catching up with B, which is moving in the same direction with speed u_B. After the collision, either B moves away from A with speed v_B (where $v_B > v_A$) or A and B continue together (in the case that $v_B = v_A$). The situation is illustrated in Figure 7.2.

▲ **Figure 7.2**

The law of conservation of momentum now gives

momentum after collision = momentum before collision

$$m_A\,v_A + m_B\,v_B = m_A\,u_A + m_B\,u_B$$

Example 7.4

Ball A of mass $0.5\,\text{kg}$ travelling at $3\,\text{m s}^{-1}$ hits stationary ball B of mass $0.2\,\text{kg}$.

After the collision, ball A is stationary.

(i)   Draw diagrams showing the situation before and after the collision.

(ii) Find the speed of ball B after the collision.

(iii) Find the impulse on each ball.

→

Solution

(i)

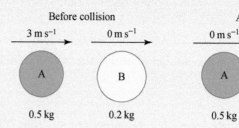

Before collision   After collision

▲ Figure 7.3

(ii) Using conservation of momentum

$$0.5 \times 0 + 0.2v = 0.5 \times 3 + 0.2 \times 0$$

$$0.2v = 1.5$$

$$v = 7.5\,\mathrm{ms^{-1}}$$

B moves away with speed $7.5\,\mathrm{ms^{-1}}$.

(iii) Impulse = final momentum − initial momentum

Impulse on ball A $= 0.5 \times 0 - 0.5 \times 3 = -1.5\,\mathrm{Ns}$

Impulse on ball B $= 0.2 \times 7.5 - 0.2 \times 0 = 1.5\,\mathrm{Ns}$

Exercise 7B

1 A van of mass $2800\,\mathrm{kg}$ travelling at $10\,\mathrm{ms^{-1}}$ collides head-on with a car of mass $800\,\mathrm{kg}$ that is travelling in the opposite direction at $25\,\mathrm{ms^{-1}}$. After the collision, the van comes to rest. Taking the direction of the van's motion before collision as the positive direction, find

(i) the velocity of the car

(ii) the impulse on each vehicle.

2 A railway truck of mass 25 tonnes is shunted with speed $4\,\mathrm{ms^{-1}}$ towards a stationary truck of mass 20 tonnes. What is the speed of the 25 tonne truck after impact

(i) if the two trucks remain in contact

(ii) if the 20 tonne truck now moves at $4\,\mathrm{ms^{-1}}$?

3 A spaceship of mass $25\,000\,\mathrm{kg}$ travelling with speed $1500\,\mathrm{ms^{-1}}$ docks with a space station of mass $600\,000\,\mathrm{kg}$ travelling at $1475\,\mathrm{ms^{-1}}$ in the same direction.

What is the speed of the combined spaceship and space station after docking is completed?

4 A rifle of mass $5\,\mathrm{kg}$ is used to fire a bullet of mass $50\,\mathrm{grams}$ at a speed of $250\,\mathrm{ms^{-1}}$. Calculate the recoil speed of the gun.

5 A child of mass $35\,\mathrm{kg}$, who is running through a supermarket at $5\,\mathrm{ms^{-1}}$, leaps on to a stationary shopping trolley of mass $12\,\mathrm{kg}$. Find the speed at which the child and trolley move off together, assuming that the trolley is free to move easily.

7.3 Newton's experimental law

If you drop two different balls, say a tennis ball and a cricket ball, from the same height, will they both rebound to the same height as they were dropped from? If not, will they rebound to the same height as each other? How will the heights of the second bounces compare with the heights of the first ones?

Your own experience probably tells you that different balls will rebound to different heights.

For example, a tennis ball will rebound to a greater height than a cricket ball. Also, the surface on which the ball is dropped will affect the height of the bounce. A tennis ball dropped onto a concrete floor will rebound higher than when it is dropped onto a carpeted floor. The following experiment allows you to look at this situation more closely.

EXPERIMENT

The aim of this experiment is to investigate what happens when balls bounce.

Draw up a table to record your results.

1   Drop a ball from a variety of heights and record the heights of release h_a and rebound h_s. Repeat several times for each height.
2   Use your values of h_a and h_s to calculate v_a and v_s, the speeds on impact and rebound. Enter your results in your table.
3   Calculate the ratio $\frac{v_a}{v_s}$ for each pair of readings of h_a and h_s and enter the results in your table.
4   What do you notice about these ratios?
5   Repeat the experiment with different types of ball.

Coefficient of restitution

Newton's experiments on collisions led him to formulate a simple law relating to the speeds before and after a direct collision between two bodies, called **Newton's experimental law**. It is also known as **Newton's law of restitution** or **Newton's law of impact**.

$$\frac{\text{speed of separation}}{\text{speed of approach}} = \text{constant}$$

This can be written as

speed of separation = constant × speed of approach

This constant is called the **coefficient of restitution** and is conventionally denoted by the letter e. For two particular surfaces, e is a constant between 0 and 1. It does not have any units, since it is the ratio of two speeds.

For very bouncy balls, e is close to 1, and for balls that do not bounce, e is close to 0. A collision for which $e = 1$ is called perfectly elastic, and a collision for which $e = 0$ is called perfectly inelastic.

For perfectly elastic collisions there is no energy loss. For perfectly inelastic collisions the objects coalesce and the energy loss is the largest it can be.

Direct impact with a fixed surface

The value of e for the ball you used in the experiment is given by $\dfrac{v_s}{v_a}$, and your experiment should have proved that, for a given ball, this had approximately the same value each time. When a moving object hits a fixed surface that is perpendicular to its motion, it rebounds in the opposite direction. If the speed of approach is v_a and the speed of separation is v_s,

Newton's experimental law gives

$$\frac{v_s}{v_a} = e$$

$$\Rightarrow \quad v_s = e v_a$$

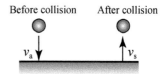

▲ Figure 7.4

Collisions between bodies moving in the same straight line

Figure 7.5 shows two objects that collide while moving along a straight line. Object A is catching up with B, and after the collision either B moves away from A or they continue together.

Before   +   After

u_A u_B v_A v_B

m_A m_B m_A m_B

▲ Figure 7.5

$u_A > u_B$ for the collision to occur.

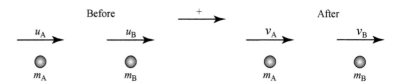

Speed of approach: $u_A - u_B$

Speed of separation: $v_B - v_A$

$v_B \geqslant v_A$ as B moves away from A, or if the particles coalesce then $v_A = v_B$.

By Newton's experimental law

$$\text{speed of separation} = e \times \text{speed of approach}$$

$$\Rightarrow \qquad v_B - v_A = e(u_A - u_B)$$

| Example 7.5 | A direct collision takes place between two snooker balls. The white cue ball travelling at $2\,\text{m s}^{-1}$ hits a stationary red ball. After the collision, the red ball moves in the direction in which the cue ball was moving before the collision. The balls have equal mass and the coefficient of restitution between the two balls is 0.6. Predict the velocities of the two balls after the collision. |

Solution

Let the mass of each ball be m. Before the collision, their velocities are u_W and u_R. After the collision, their velocities are v_W and v_R.

The situation is summarised in Figure 7.6.

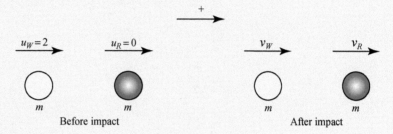

▲ **Figure 7.6**

Speed of approach $= 2 - 0 = 2$

Speed of separation $= v_R - v_W$

Speed of separation $= e \times$ Speed of approach

By Newton's experimental law

$$\Rightarrow \quad v_R - v_W = 0.6 \times 2$$

$$\Rightarrow \quad v_R - v_W = 1.2 \qquad\qquad ①$$

Conservation of momentum

$$mv_W + mv_R = mu_W + mu_R$$

Dividing through by m, and substituting $u_W = 2$, $u_R = 0$, this becomes

$$v_W + v_R = 2 \qquad\qquad ②$$

Adding ① + ② gives $2v_R = 3.2$

so $v_R = 1.6$ and, from equation ②, $v_W = 0.4$.

After the collision both balls move in the original direction of the white cue ball, the red ball at a speed of $1.6\,\text{m s}^{-1}$ and the white cue ball at a speed of $0.4\,\text{m s}^{-1}$.

Example 7.6

An object A of mass m moving with speed $2u$ hits an object B of mass $2m$ moving with speed u in the opposite direction to A. The coefficient of restitution is e.

(i) Show that the ratio of speeds remains unchanged whatever the value of e.

(ii) Find the loss of kinetic energy in terms of m, u and e.

Solution

(i) Let the velocities of A and B after the collision be v_A and v_B, respectively.

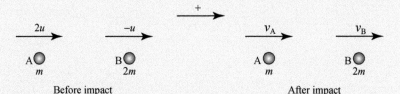

$+$

| $2u$ | $-u$ | | v_A | v_B |

A$\bigcirc$ B$\bigcirc$ A$\bigcirc$ B$\bigcirc$
m $2m$ m $2m$

Before impact   After impact

▲ **Figure 7.7**

Speed of approach $= 2u - (-u) = 3u$

Speed of separation $= v_B - v_A$

Using Newton's experimental law

Speed of separation $= e \times$ speed of approach

$$v_B - v_A = e \times 3u \qquad \textcircled{1}$$

Conservation of momentum gives

$$mv_A + 2mv_B = m(2u) + 2m(-u)$$

Dividing by m gives

$$v_A + 2v_B = 0 \qquad \textcircled{2}$$

Equation $\textcircled{1}$ is $v_B - v_A = 3eu$

Adding $\textcircled{1} + \textcircled{2}$ $3v_B = 3eu$

$$v_B = eu$$

From $\textcircled{2}$ $v_A = -2eu$

The ratio of speeds was initially $2u : u$ and finally $2eu : eu$, so the ratio of speeds is unchanged at $2 : 1$ (providing $e \neq 0$).

(ii) Initial K.E. of A $\frac{1}{2}m \times (2u)^2 = 2mu^2$

Initial K.E. of B $\frac{1}{2}(2m) \times u^2 = mu^2$

Total K.E before impact $= 3mu^2$

Final K.E. of A $\frac{1}{2}m \times 4e^2u^2 = 2me^2u^2$

Final K.E. of B $\frac{1}{2}(2m) \times e^2u^2 = me^2u^2$

Total K.E. after impact $= 3me^2u^2$

Loss of K.E. $= 3mu^2(1 - e^2)$

Note

In this example, A and B lose *all* their energy when $e = 0$, but this is not true in general.

Only when $e = 1$ is there no loss in K.E. Kinetic energy is lost in any collision in which the coefficient of restitution is not equal to 1.

Exercise 7C

You will find it helpful to draw diagrams when answering these questions.

1   In each of these situations, find the unknown quantity, which may be the initial speed u, the final speed v or the coefficient of restitution e.

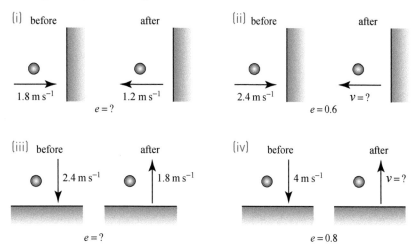

2   Find the coefficient of restitution in each situation.

(i)   A football hits the goalpost at $10\,\text{m s}^{-1}$ and rebounds in the opposite direction with speed $3\,\text{m s}^{-1}$.

(ii) A beanbag is thrown against a wall with speed $5\,\text{m s}^{-1}$ and falls straight down to the ground.

(iii) A bouncy ball is dropped onto the ground, landing with speed $8\,\text{m s}^{-1}$, and rebounds with speed $7.6\,\text{m s}^{-1}$.

(iv) A photon approaches a mirror along a line normal to its surface with speed $3 \times 10^8\,\text{m s}^{-1}$ and leaves it along the same line with speed $3 \times 10^8\,\text{m s}^{-1}$.

3   A tennis ball of mass $60\,\text{g}$ is hit against a practice wall. At the moment of impact it is travelling horizontally with speed $15\,\text{m s}^{-1}$. Just after the impact its speed is $12\,\text{m s}^{-1}$, also horizontally. Find

(i)   the coefficient of restitution between the ball and the wall

(ii) the impulse acting on the ball

(iii) the loss of kinetic energy during the impact.

PS 4   A ball of mass $80\,\text{g}$ is dropped from a height of $1\,\text{m}$ onto a level floor and bounces back to a height of $0.81\,\text{m}$. Find

(i)   the speed of the ball just before it hits the floor

(ii) the speed of the ball just after it has hit the floor

(iii) the coefficient of restitution

(iv) the change in the kinetic energy of the ball from just before it hits the floor to just after it leaves the floor

(v)   the change in the potential energy of the ball from the moment when it was dropped to the moment when it reaches the top of its first bounce

(vi) the height of the ball's next bounce.

Answers to exercises are available at www.hoddereducation.com/cambridgeextras

5   In each of these situations, a collision is about to occur. Masses are given in kilograms, speeds are in metres per second. In each case

(i)   draw diagrams showing the situation before and after impact, including known velocities and the symbols you are using for velocities that are not yet known

(ii) use the equations corresponding to the law of conservation of momentum and to Newton's experimental law to find the final velocities

(iii) find the loss of kinetic energy during the collision.

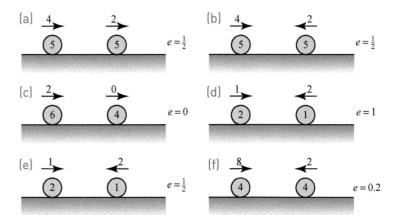

6   Two children drive dodgems straight at each other, and collide head-on. Both dodgems have the same mass (including their drivers) of $150\,\text{kg}$. Reeta is driving at $3\,\text{m}\,\text{s}^{-1}$, Atul at $2\,\text{m}\,\text{s}^{-1}$. After the collision Reeta is stationary. Find

(i)   Atul's velocity after the collision

(ii) the coefficient of restitution between the cars

(iii) the impulse acting on Atul's car

(iv) the kinetic energy lost in the collision.

7   A trapeze artist of mass $50\,\text{kg}$ falls from a height of $20\,\text{m}$ into a safety net.

(i)   Find the speed with which she hits the net. (You may ignore air resistance.)

Her speed on leaving the net is $15\,\text{m}\,\text{s}^{-1}$.

(ii) What is the coefficient of restitution between her and the net?

(iii) What impulse does the trapeze artist receive?

(iv) How much mechanical energy is absorbed in the impact?

(v)   If you were a trapeze artist, would you prefer a safety net with a high coefficient of restitution or a low one? Justify your answer.

8   Two spheres of equal mass, m, are travelling towards each other along the same straight line when they collide. Both have speed v just before the collision and the coefficient of restitution between them is e. Your answers should be given in terms of m, v and e.

   (i)   Draw diagrams to show the situation before and after the collision.

   (ii)   Find the velocities of the spheres after the collision.

   (iii) Show that the kinetic energy lost in the collision is given by $mv^2(1 - e^2)$.

   (iv)   Explain why the result in part (iii) shows that e cannot have a value greater than 1.

9   Three identical spheres are lying in the same straight line. The coefficient of restitution between any pair of spheres is $\frac{1}{2}$. Initially the left-hand sphere has a velocity of $2\,\mathrm{m\,s^{-1}}$ towards the other two, which are both stationary. What are the final velocities of all three, when no more collisions can occur?

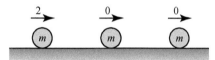

10 The diagram shows two snooker balls and one edge cushion. The coefficient of restitution between the balls and the cushion is 0.5 and that between the balls is 0.75. Ball A (the cue ball) is hit directly towards the stationary ball B with speed $8\,\mathrm{m\,s^{-1}}$. Find the speed and directions of the two balls after their second impact with each other.

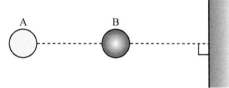

CP

11 The coefficient of restitution between a ball and the floor is e. The ball is dropped from a height h. Air resistance may be neglected, and your answers should be given in terms of e, h, g and n, the number of bounces.

   (i)   Find the time it takes the ball to reach the ground and its speed when it arrives there.

   (ii)   Find the ball's height at the top of its first bounce.

   (iii) Find the height of the ball at the top of its nth bounce.

   (iv)   Find the time that has elapsed when the ball hits the ground for the second time, and for the nth time.

   (v)   Show that, according to this model, the ball comes to rest within a finite time having completed an infinite number of bounces.

   (vi)   What distance does the ball travel before coming to rest?

Answers to exercises are available at www.hoddereducation.com/cambridgeextras

CP 12 Three identical smooth spheres A, B and C lie at rest on a smooth horizontal table with their centres in a straight line and B lying between A and C. A is projected towards B with speed u. If the coefficient of restitution at each collision is e, where $0 < e < 1$, find the velocity of each of the spheres just after C is set in motion. Show that A strikes B a second time whatever the value of e and that B strikes C a second time if e is less than $3 - 2\sqrt{2}$.

7.4 Oblique impact of a sphere on a plane

When an object hits a smooth plane there can be no impulse parallel to the plane so the component of momentum, and hence velocity, is unchanged in this direction. Perpendicular to the plane, the momentum is changed but Newton's experimental law still applies.

Figure 7.8 shows the components of the velocity of a ball immediately before and after it hits a smooth plane with coefficient of restitution e.

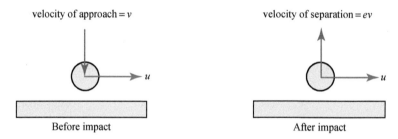

velocity of approach $= v$ velocity of separation $= ev$

Before impact   After impact

▲ **Figure 7.8**

When the ball is travelling with speed U at an angle α to the plane, the components of the final velocity are $U \cos \alpha$ parallel to the plane and $eU \sin \alpha$, perpendicular to the plane.

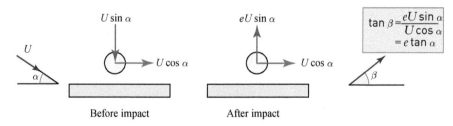

$$\tan \beta = \frac{eU \sin \alpha}{U \cos \alpha}$$
$$= e \tan \alpha$$

Before impact   After impact

▲ **Figure 7.9**

Impulse on the ball = final momentum − initial momentum. Impulse acts perpendicular to the plane because there is no change in momentum parallel to the plane.

In Figure 7.8 the impulse is:

$$mev - m(-v) = (1 + e)mv \text{ upwards.}$$

In Figure 7.9 the impulse is:

$$meU \sin \alpha - m(-U \sin \alpha) = (1 + e)mU \sin \alpha \text{ upwards.}$$

Whenever an impact takes place, energy is likely to be lost. In the cases illustrated in the diagrams, the loss in kinetic energy is:

$$\tfrac{1}{2}m(u^2 + v^2) - \tfrac{1}{2}m(u^2 + e^2 v^2) = \tfrac{1}{2}m(1 - e^2)v^2$$

or $\qquad \tfrac{1}{2}m(1 - e^2)U^2 \sin^2 \alpha.$

> ❓ What happens to the ball when $e = 0$ and $e = 1$?

Example 7.7

A ball of mass $0.2\,\text{kg}$ moving at $12\,\text{ms}^{-1}$ hits a smooth horizontal plane at an angle of $75°$ to the horizontal. The coefficient of restitution is 0.5. Find

(i)   the impulse on the ball

(ii) the impulse on the plane

(iii) the kinetic energy lost by the ball.

Solution

(i) Figure 7.10 shows the velocities before and after impact.

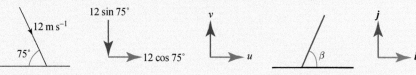

▲ **Figure 7.10**

Parallel to the plane: $u = 12 \cos 75°$ ← No change in velocity parallel to the plane.

Perpendicular to the plane: $v = 0.5 \times 12 \sin 75°$ ← Using Newton's experimental law with $e = 0.5$.

$$= 6 \sin 75°$$

The impulse on the ball = final momentum − initial momentum

$$\mathbf{J} = 0.2 \begin{pmatrix} 12 \cos 75° \\ 6 \sin 75° \end{pmatrix} - 0.2 \begin{pmatrix} 12 \cos 75° \\ -12 \sin 75° \end{pmatrix}$$ ← Using directions $\mathbf{i}$ and $\mathbf{j}$ as shown.

$$= \begin{pmatrix} 0 \\ 3.6 \sin 75° \end{pmatrix}$$

The impulse on the ball is $3.6 \sin 75° \, \mathbf{j}$, that is, $3.48\,\text{Ns}$ perpendicular to the plane and upwards in the $\mathbf{j}$ direction.

(ii) By Newton's third law, the impulse on the plane is equal and opposite to the impulse on the ball. It is $3.48\,\text{Ns}$ perpendicular to the plane in the direction of $-\mathbf{j}$.

→

Answers to exercises are available at www.hoddereducation.com/cambridgeextras

(iii) The initial kinetic energy $= \frac{1}{2} \times 0.2 \times 12^2 = 14.4 \, \text{J}$

Final kinetic energy $= \frac{1}{2} \times 0.2 \times \left[(12\cos 75°)^2 + (6\sin 75°)^2 \right]$

$= 4.32... \, \text{J}$

Loss in kinetic energy $= 14.4 - 4.32... = 10.1 \, \text{J}$

Example 7.8

A ball moving with speed $10 \, \text{ms}^{-1}$ hits a smooth horizontal plane at an angle of 60° to the horizontal. The coefficient of restitution between the ball and the surface is $\frac{1}{3}$. The ball rebounds with speed v at an angle β with the surface.

(i) Find v.

(ii) Find β.

Solution

(i)

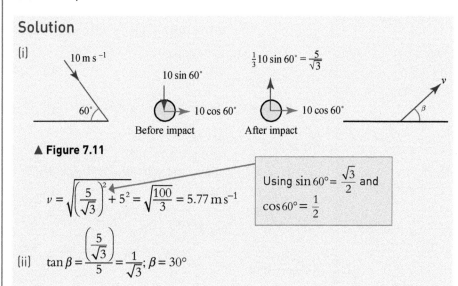

▲ **Figure 7.11**

$$v = \sqrt{\left(\frac{5}{\sqrt{3}} \right)^2 + 5^2} = \sqrt{\frac{100}{3}} = 5.77 \, \text{ms}^{-1}$$

Using $\sin 60° = \frac{\sqrt{3}}{2}$ and $\cos 60° = \frac{1}{2}$

(ii) $\tan \beta = \dfrac{\left(\frac{5}{\sqrt{3}} \right)}{5} = \dfrac{1}{\sqrt{3}}; \beta = 30°$

Exercise 7D

1 In each case, find the velocity of the object after one impact with a smooth plane. Give your answer as a vector showing the horizontal and vertical components of the velocity.

(i) Initial velocity $4 \, \text{ms}^{-1}$ at 20° to the plane.
Coefficient of restitution 0.5.

(ii) Initial velocity $10 \, \text{ms}^{-1}$ at 40° to the plane.
Coefficient of restitution 0.1.

(iii) Initial velocity $u \, \text{ms}^{-1}$ at $\alpha°$ to the plane.
Coefficient of restitution 0.8.

2 A ball of mass $0.1 \, \text{kg}$, moving at $10 \, \text{ms}^{-1}$, hits a smooth horizontal plane at an angle of 80° to the horizontal. The coefficient of restitution is 0.6. Taking horizontal and vertical unit vectors **i** and **j** respectively, find

(i) the impulse on the ball and the direction in which it acts

(ii) the impulse on the plane and the direction in which it acts

(iii) the kinetic energy lost by the particle.

3   A particle of mass $0.05\,\text{kg}$, moving at $8\,\text{m s}^{-1}$, hits a smooth horizontal plane at an angle of $45°$ to the horizontal. The coefficient of restitution is 0.6. Taking horizontal and vertical unit vectors **i** and **j** respectively, find

(i)   the impulse on the particle and the direction in which it acts

(ii)   the impulse on the plane and the direction in which it acts

(iii)   the kinetic energy lost by the particle.

4   A ball of mass $m\,\text{kg}$, moving at $u\,\text{m s}^{-1}$, hits a smooth horizontal plane at an angle of $\alpha°$ to the horizontal. The coefficient of restitution is 0.

(i)   Calculate the impulse on the ball.

(ii)   Show that the kinetic energy lost is $\frac{1}{2}mu^2\sin^2\alpha$.

5   Show that the kinetic energy lost by a particle of mass $m\,\text{kg}$ that hits a smooth plane when it is moving with velocity $u\,\text{m s}^{-1}$ at an angle of $\alpha°$ to the plane, is $\frac{1}{2}mu^2\left(1-e^2\right)\sin^2\alpha$, where e is the coefficient of restitution.

CP   6   A ball is hit from level ground with initial components of velocity $u_x\,\text{m s}^{-1}$ horizontally and u_y vertically. Assume the ball is a particle and ignore air resistance.

(i)   Show that its horizontal range is $R = \dfrac{2u_x u_y}{g}$.

The ball bounces on the ground with coefficient of restitution 0.6.

(ii)   Find an expression, in terms of R, for the horizontal distance the ball travels between the first and second bounce.

(iii)   Find an expression, in terms of R and n, for the horizontal distance the ball travels between the nth and $(n + 1)$th bounce.

(iv)   By considering the sum of a geometric series, calculate the total horizontal distance travelled by the ball up to the sixth bounce.

7   A ball of mass $0.5\,\text{kg}$ falls from a height of $62.5\,\text{cm}$ onto a plane inclined at $60°$ to the horizontal.

The coefficient of restitution between the ball and the plane is $\frac{4}{7}$. Find

(i)   the speed of the ball immediately before striking the plane

(ii)   the magnitude and direction of the velocity of the ball after impact

(iii)   the loss in kinetic energy of the ball due to the collision.

8   A small marble is projected horizontally over the edge of a table $0.8\,\text{m}$ high at a speed of $2.5\,\text{m s}^{-1}$ and bounces on smooth horizontal ground with coefficient of restitution 0.7.

Calculate

(i)   the components of the velocity of the marble just before it hits the ground

(ii)   its horizontal distance from the edge of the table when it first hits the ground

(iii) the horizontal distance travelled between the first and second bounces

(iv) the horizontal distance travelled between the nth and $(n + 1)$th bounces

(v) the number of bounces before the distance between bounces is less than 20 cm.

M 9 A smooth snooker ball moving at $2\,\mathrm{m\,s^{-1}}$ hits a cushion at an angle of 30° to the cushion.

The ball then rebounds and hits a second cushion that is perpendicular to the first. The coefficient of restitution for both impacts is 0.8.

(i) Find the direction of motion after each impact.

(ii) Find the magnitude of the velocity after the second impact.

(iii) Repeat parts (i) and (ii) for a ball moving at $u\,\mathrm{m\,s^{-1}}$ that hits the first cushion at an angle α. Assume the coefficient of restitution is e. Hence show that the direction of a ball is always reversed after hitting two perpendicular cushions and state the factor by which its speed is reduced.

CP 10 A ball falls vertically and strikes a fixed plane inclined at an angle θ $(\theta < 45°)$ to the horizontal. The coefficient of restitution is $\dfrac{7}{25}$ and the ball rebounds horizontally. Show that

(i) $\tan\theta = \dfrac{1}{5}\sqrt{7}$

(ii) the fraction of kinetic energy lost in the collision is $\dfrac{18}{25}$.

7.5 Oblique impact of smooth elastic spheres

Two smooth spheres, A of mass m_A and B of mass m_B, collide. Immediately before impact the velocity of A is u at an angle α with the line of centres of the spheres, and the velocity of B is v at an angle β with the line of centres.

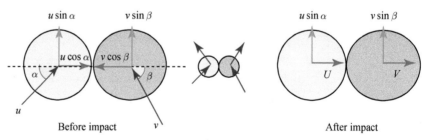

Before impact   After impact

▲ **Figure 7.12**

When analysing an impact like this you need to consider the components of the motion in two directions: parallel to the line joining the centres of A and B, and perpendicular to the line joining these centres.

Motion perpendicular to the line of centres

There is no impulse between the spheres in the direction of their common tangent at the point of contact, and the momentum of each sphere in this direction is unchanged by the impact.

Hence the component velocities after the collision in this direction are $u \sin \alpha$ for A and $v \sin \beta$ for B.

Motion along the line of centres

Considering the components of the velocities parallel to the line joining A and B, conservation of momentum gives

$$m_A u \cos \alpha - m_B v \cos \beta = m_A U + m_B V$$

The coefficient of restitution is e so that Newton's experimental law for motion in a horizontal direction gives

$$e\left(u \cos \alpha + v \cos \beta\right) = V - U \quad \longleftarrow \quad \boxed{\text{Speed of separation}}$$

$\boxed{\text{Speed of approach}}$ These two simultaneous equations are sufficient to determine U and V.

Once you know U and V, the velocities of A and B after the collision can be written in vector form as

$$\mathbf{v}_A = U\mathbf{i} + u \sin\alpha \, \mathbf{j} \quad \text{and} \quad \mathbf{v}_B = V\mathbf{i} + v \sin\beta \, \mathbf{j}$$

where the directions parallel and perpendicular to the line of centres are denoted by $\mathbf{i}$ and $\mathbf{j}$, respectively.

Example 7.9

In a game of snooker the cue ball, moving with speed $2\,\text{m s}^{-1}$, strikes a stationary red ball. The cue ball is moving at an angle of $60°$ to the line of centres of the two balls. Both balls are smooth and have the same mass m. The coefficient of restitution between the balls is $\frac{1}{2}$.

Find the velocities of the two balls after impact.

Solution

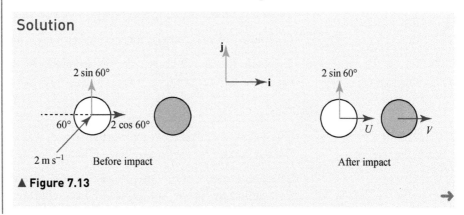

▲ Figure 7.13

Answers to exercises are available at www.hoddereducation.com/cambridgeextras

For motion after the impact, let the component velocities parallel to the line joining the centres (the direction of $\mathbf{i}$) be U for the cue ball and V for the red ball.

Conservation of momentum in the direction of $\mathbf{i}$:

$$m \times 2 \cos 60° = mU + mV$$

$$1 = U + V \quad \text{(A)} \quad \boxed{\text{Divide by } m \text{ and use } \cos 60° = 0.5.}$$

Newton's experimental law:

$$e \times 2 \cos 60° = V - U \quad \boxed{\text{Speed of separation}}$$

$$\tfrac{1}{2} = V - U \quad \text{(B)}$$

$\boxed{\text{Speed of approach}}$

(A)+(B): $\quad \tfrac{3}{2} = 2V \Rightarrow V = \tfrac{3}{4}$

(A)–(B): $\quad \tfrac{1}{2} = 2U \Rightarrow U = \tfrac{1}{4}$

The velocity of the red ball is $\tfrac{3}{4}\mathbf{i}$, the velocity of the white ball is $\tfrac{1}{4}\mathbf{i} + 2\sin 60°\mathbf{j} = \tfrac{1}{4}\mathbf{i} + \sqrt{3}\mathbf{j}$.

The red ball moves with speed $0.75\,\text{m s}^{-1}$ along the line of centres. The cue ball moves with speed $1.75\,\text{m s}^{-1}$ at $82°$ to the line of centres.

$$\boxed{\sqrt{\left(\tfrac{1}{4}\right)^2 + 3} = \sqrt{\tfrac{49}{16}} = \tfrac{7}{4}}$$

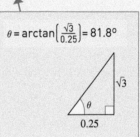

$$\theta = \arctan\left(\tfrac{\sqrt{3}}{0.25}\right) = 81.8°$$

Example 7.10

A smooth sphere A of mass $2m$, moving with speed $4\,\text{m s}^{-1}$, collides with a smooth sphere B of mass m that is moving with speed $2\,\text{m s}^{-1}$. The velocity of A immediately before impact makes an angle of $45°$ to the line of centres. The velocity of B immediately before impact is at $90°$ to the line of centres. The coefficient of restitution between the two balls is 0.6.

(i) Draw a diagram showing the situation before and after the collision.

(ii) Calculate the velocities of the two spheres after impact.

(iii) Calculate the loss of kinetic energy sustained by the system during the impact.

Solution

(i) The velocities of the two spheres are shown in Figure 7.14, as well as their components along and perpendicular to the line of centres.

The components perpendicular to the line of centres ($4\sin 45° = 2\sqrt{2}$ for A and 2 for B) are not affected by the collision.

The components along the line of centres are taken to be V_A and V_B.

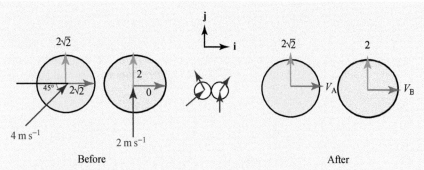

Before   After

▲ **Figure 7.14**

Conservation of momentum along line of centres:

(ii) $2m \times 4\cos45° = 2mV_A + mV_B$

$\rightarrow 4\sqrt{2} = 2V_A + V_B$ (A)

> Using $\cos45° = \frac{\sqrt{2}}{2}$ and dividing by m

Newton's experimental law:

$e \times 4\cos45° = V_B - V_A$ $2\sqrt{2}e = V_B - V_A$ (B)

(A)−(B): $(4 - 2e)\sqrt{2} = 3V_A$

$V_A = \frac{\sqrt{2}}{3}(4 - 2e)$ ← $4 - 2e = 4 - 2 \times 0.6$

$V_A = \frac{2.8\sqrt{2}}{3} = 1.31\ldots$

(A)+2(B): $4(1 + e)\sqrt{2} = 3V_B$

> $\sqrt{1.31\ldots^2 + (2\sqrt{2})^2}$
> $= \sqrt{1.31\ldots^2 + 8} = 3.12$

$V_B = \frac{4\sqrt{2}}{3}(1 + e)$ ← $4(1 + e) = 4(1 + 0.6)$

$V_B = \frac{6.4\sqrt{2}}{3} = 3.01\ldots$

> $\theta = \arctan\left(\frac{2\sqrt{2}}{1.31\ldots}\right) = 65.0°$
>
> $2\sqrt{2}$
>
> θ
>
> $1.31\ldots$

Velocity of A after impact: $1.31\ldots\mathbf{i} + 2\sqrt{2}$

A moves with speed $3.12\,\text{m s}^{-1}$ at an angle of $65.0°$ to the line of centres.

Velocity of B after impact: $3.01\ldots\mathbf{i} + 2\mathbf{j}$

B moves with speed $3.62\,\text{m s}^{-1}$ at an angle of $33.5°$ to the line of centres.

> $\sqrt{3.01\ldots^2 + 2^2} = 3.62$
>
> $\phi = \arctan\left(\frac{2}{3.01\ldots}\right) = 33.5°$
>
> 2
>
> ϕ
>
> $3.01\ldots$

(iii) The kinetic energy of the system before the collision is equal to:

$\frac{1}{2} \times 2m \times 4^2 + \frac{1}{2} \times m \times 2^2 = 18m$

The kinetic energy of the system after the collision is equal to:

$\frac{1}{2} \times 2m \times 3.12^2 + \frac{1}{2} \times m \times 3.62^2 = 16.29m$

The loss in kinetic energy is:

$18m - 16.29m = 1.71m$

Answers to exercises are available at www.hoddereducation.com/cambridgeextras

Note

This result could have been obtained by consideration of the contribution from the components of velocity along the line of centres only, as there is no change arising from the components perpendicular to the line of centres:

K.E. before: $\frac{1}{2} \times 2m \times \left(2\sqrt{2}\right)^2 + \frac{1}{2} \times m \times 0 = 8m$

K.E. after: $\frac{1}{2} \times 2m \times \frac{2.8\sqrt{2}^2}{3} + \frac{1}{2} \times m \times \frac{6.4\sqrt{2}^2}{3} = 6.29...m$

Loss in K.E.: $8m - 6.29...m = 1.71m$

Exercise 7E

In questions **1–6**, a smooth sphere A of mass m_A collides with a smooth sphere B of mass m_B, as shown in the diagram.

The coefficient of restitution between the spheres is e.

Immediately before the collision, A is moving with speed u_A at an angle α with the line of centres and B is moving with speed u_B at an angle β with the line of centres.

Immediately after impact, A is moving with speed v_A at an angle α_A with the line of centres and B is moving with speed v_B at an angle β_B with the line of centres.

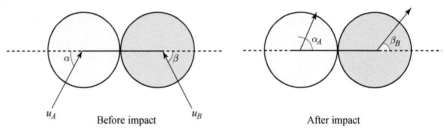

Before impact   After impact

1   $m_A = 4\,\text{kg}$, $m_B = 2\,\text{kg}$, $u_A = 2\,\text{ms}^{-1}$, $u_B = 4\,\text{ms}^{-1}$, $\alpha = \beta = 45°$, $e = 0.5$.
   Calculate v_A, v_B, α_A and β_B.

2   $m_A = m_B = m$, $\alpha = 60°$, $u_A = u$, $u_B = 0$, $e = 0.6$.
   Calculate v_A, v_B, α_A and β_B.

3   $m_A = m_B = m$, $u_A = u_B = u$, $\alpha = 0°$, $\beta = 90°$, $e = 0.5$.
   Calculate v_A, v_B, α_A and β_B.

4   $m_A = m_B = m$, $u_A = u_B = u$, $\alpha = 60°$, $\beta = 60°$, $e = 0.5$.
   Calculate v_A, v_B, α_A and β_B.

5   $m_A = m$, $m_B = 5m$, $u_A = u$, $u_B = 0$, $\alpha = 60°$, $\alpha_A = 90°$.
   Calculate e.

6   $m_A = m_B = m$, $u_B = 0$, $\alpha = 45°$, $e = \frac{2}{3}$.
   Calculate α_A.

7 Two identical smooth balls of mass m are moving with equal speed u in opposite directions. The balls collide obliquely, so that the line of centres between the balls is at 30° to the direction of motion. Show that the loss in kinetic energy due to the impact is 75% of what it would be if the impact were direct.

(PS) **8** A smooth sphere A of mass $2m$ moving with speed $2u$ collides with a smooth sphere B of mass m moving with speed u.

At the moment of impact, A is moving at 60° to the line of centres and B is moving at 90° to the line of centres.

The coefficient of restitution between the spheres is 0.5. Find

(i) the component of velocity along the line of centres after impact for each sphere

(ii) the speed of each sphere after impact and the angle each makes with the line of centres

(iii) the loss in kinetic energy for the system.

(CP) **9** In this question all the discs are circular and have the same radius.

(i) A disc of mass m is sliding across a table when it collides with a stationary disc with the same mass. After the collision, the directions of motion of the two discs are at right angles. Prove that the collision is perfectly elastic.

(ii) On another occasion the disc of mass m collides with a stationary disc of mass km, where $k > 1$, and the directions of their subsequent motion are at right angles. The coefficient of restitution is e. Prove that $e = \frac{1}{k}$.

(iii) State a modelling assumption required for parts (i) and (ii).

10 The diagram illustrates a collision between two smooth spheres of equal mass m. Initially they are moving along parallel lines but in opposite directions. At impact the acute angle between their line of centres and the directions of their original movement is α. The coefficient of restitution in the collision is e. Before the impact both spheres have speed u.

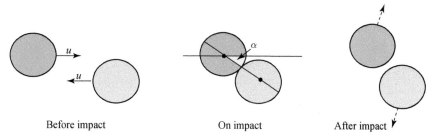

Before impact   On impact   After impact

(i) Show that the loss of kinetic energy in the collision is $mu^2 \cos^2 \alpha \left(1 - e^2\right)$.

(ii) Show that, in the case when $\alpha = 30°$ and $e = \frac{1}{3}$, the direction of motion of each of the spheres after impact is at right angles to its direction before impact.

Answers to exercises are available at www.hoddereducation.com/cambridgeextras

KEY POINTS

1 The impulse from a force **F** is given by **F**t where t is the time for which the force acts.

2 Impulse is conventionally denoted by **J**. It is a vector quantity.

3 The momentum of a body of mass m travelling with velocity **v** is given by m**v**. Momentum is a vector quantity.

4 The S.I. unit of impulse and momentum is the newton second (N s).

5 The impulse–momentum equation is

impulse = final momentum − initial momentum.

6 The law of conservation of momentum states that when no external forces are acting on a system, the total momentum of the system is constant. Since momentum is a vector quantity, this applies to the magnitude of the momentum in any direction.

7 Newton's experimental law:

$$\text{Coefficient of restitution, } e = \frac{\text{speed of separation}}{\text{speed of approach}}$$

speed of separation = $e \times$ speed of approach

8 Collision between a sphere and a fixed plane

Component of velocity parallel to surface remains unchanged
[$v \cos \beta = u \cos \alpha$]

Component of velocity perpendicular to surface: [$v \sin \beta = -eu \sin \alpha$]

Loss in kinetic energy: $\frac{1}{2}mu^2 \sin^2 \alpha (1-e^2)$

9 Oblique impact between smooth spheres

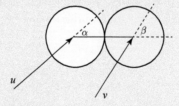 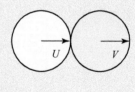

Perpendicular to line of centres
$u \sin \alpha$ and $v \sin \beta$ remain unchanged by the collision.

Along line of centres

Conservation of momentum:

$$m_A u \cos\alpha + m_B v \cos\beta = m_A U + m_B V \qquad \text{①}$$

Newton's experimental law:

$$V - U = e(u \cos\alpha - v \cos\beta) \qquad \text{②}$$

Equations ① and ② can be solved to find U and V.

LEARNING OUTCOMES

Now that you have finished this chapter, you should be able to

- understand how to apply the principle of conservation of momentum to direct impacts

- understand Newton's experimental law and know the meaning of coefficient of restitution

- understand and use the fact that $0 \leqslant e \leqslant 1$

- understand the implications of values of 0 and 1 for the coefficient of restitution

- understand that when the coefficient of restitution is less than 1, energy is not conserved during an impact

- find the loss of kinetic energy during a direct impact

- understand that for perfectly elastic collisions there is no energy loss

- understand that for perfectly inelastic collisions, the energy loss is the largest it can be

- understand the term oblique impact and the assumptions made when modelling oblique impact

- understand the meaning of Newton's experimental law and of the coefficient of restitution when applied to an oblique impact

- solve problems involving impact between an object and a fixed smooth plane by considering components of motion parallel and perpendicular to the line of impulse

- solve problems involving impact between two spheres by considering components of motion in directions parallel and perpendicular to the line of centres

- calculate the loss of kinetic energy in an oblique impact.

Index